In the Service of Scholarship

Harold in his favorite environment, 1978. (Photo, B. A. King)

In the Service of Scholarship

Harold Hugo &
The Meriden Gravure Company

William J. Glick

Oak Knoll Press
New Castle, Delaware
2017

Published by
Oak Knoll Press
310 Delaware Street
New Castle, DE 19720
ISBN: 978-1-58456-356-3

Also by the author:
William Edwin Rudge, New York, 1984, The Typophiles
Typophiles Chap Book 57

Typesetting & design: Scott Vile at the Ascensius Press.

All inquiries should be addressed to Oak Knoll Press,
310 Delaware Street, New Castle, DE 19720.

Printed in the United States of America on acid-free paper meeting the requirements of ANSI/NISO Z39.48.1992 (Permanence of Paper)

Library of Congress Cataloging-in-Publication Data
Names: Glick, William J., author.
Title: In the service of scholarship : Harold Hugo & the Meriden Gravure Company / by William J. Glick.
Description: First edition. | New Castle, Delaware : Oak Knoll Press, 2017. | Includes bibliographical references and index.
Identifiers: LCCN 2016059737 | ISBN 9781584563563 (acid-free paper)
Subjects: LCSH: Hugo, Harold. | Meriden Gravure Company--History. | Printers--United States--Biography. |
Printing--Connecticut--Meriden--History--20th century. | Printing industry--Connecticut--Meriden--History--20th century. | Collotype. | Offset printing.

Contents

Preface

THERE ARE already modest amounts of information in print on the subject of Harold Hugo and the Meriden Gravure Co. However, they tend to be devoted to only one aspect of the firm's activities. Further, they appear for the most part only as privately printed limited editions for noncommercial distribution, and they rarely turn up today in the antiquarian book trade.

An important motivation for me was to make the story of Harold Hugo and the Meriden Gravure Co. more widely known to present and future generations of practitioners, students, historians and collectors of the graphic arts. My qualification for this attempt to record this episode in printing history is that I had day-to-day personal contact with Harold over many years. I joined the firm in 1961 and was assigned a wide variety of tasks under him. A great deal of my time was spent in sales and customer service, but I also did production supervision, quality control, estimating, purchasing, contract administration and billing. Harold was, until his death, my most important career mentor.

Much of what is contained here is memoir and comes from the lore and oral tradition that was passed down to me during my time at Meriden. This information has been supplemented by Gay Walker's research on the Meriden Gravure Co. and her extensive interviews during Harold's lifetime. I have drawn on these previously published sources for much useful information in my account of the entire 100-year history of the firm. This has been augmented by documents from the Hugo family and contributions from various individuals and institutions that figure in the narration. I have necessarily had to be selective and representative rather than comprehensive and all-inclusive in my recording of people, publications and events. I have included some historical background to give context to my observations. In my text I have given considerable attention to many of the important publications that made the company's reputation. John Peckham preceded me in this endeavor when he gave a talk to the Club of Odd Volumes in 1984 entitled "Adventures in Printing: A Talk on the Career of Harold Hugo." It was a selection of a baker's dozen of the most challenging and important publications produced at Meriden. His talk appeared in print in 1995, handsomely produced and issued by the Stinehour Press. It was offered to subscribing members of a half-dozen organizations of which Harold had been a member. With some exceptions, I have chosen not to dwell upon these books in my text since Peckham has already discussed them in excellent detail. When I do mention them, it is in a somewhat different context. The thirteen titles are an important part of Harold's lifetime achievement, and they are listed in the Appendix.

The span of Harold's career was a period of substantial change in printing technology, the dissemination of knowledge, business practice and social conduct. When Harold began his career in the late twenties, letterpress was king. When he retired almost fifty years later, offset dominated most areas of the printing industry and the digital revolution was sending forth its first tender but disruptive shoots. Between his passing in 1985 and the present time, digital technology has become dominant in typesetting, in design and in the reproduction of illustrations. Many of the groundbreaking procedures that Harold pioneered have become obsolete. But his attitude toward work, his refusal to compromise on quality, his attention to detail, and his unwillingness to take the line of least resistance when dealing with difficulties is applicable to any technology.

Harold's leadership of the Meriden Gravure Co. took advantage of contemporary improvements in the photo-offset process and raised the bar further yet. During his time, the nation went through the Great Depression, fought a World War, and assumed the economic, political and cultural leadership of the free world. America's postwar prosperity permitted great things. Among them was a greatly expanded system of higher education. This started with the GI Bill but continued to grow long after the last of the World War II veterans had left the classroom. The arts prospered and grew as the public taste became more sophisticated. Art museums, libraries and institutions of higher learning expanded, and new ones were founded. Harold and his company, to which he devoted his life, were both beneficiaries of these developments and important contributors to them, as will be seen.

Acknowledgments

I DEEPLY appreciate the advice and information provided by those whom I here acknowledge. They have responded to my requests for information promptly and cheerfully and have given me encouragement and support to carry my work to completion. I could not have brought this publication into being without them. I have also been the beneficiary of crowd sourcing: many individuals who have volunteered small amounts of information that became cumulatively significant. The ultimate responsibility for the accuracy of the facts presented and the reasonableness of the judgments set forth is entirely mine. Philip Hofer in his publication *Baroque Book Illustration* included the statement, "No book is complete until *Error* has crept in and affixed his sly Imprimatur." I am appropriating this charming rationalization of misinformation to my own work, as it most certainly will be needed.

I owe thanks to many; first to David R. Godine and Elton W. "Toby" Hall, who gently but insistently prodded me into taking on this work. Toby, who took up the task of recording the achievements of Rocky Stinehour, was particularly generous in sharing information and research sources with me.

I am also greatly indebted to James Mooney for his wise and experienced counsel on many editorial matters large and small; to Gay Walker for her encouragement and support, for her extensive knowledge of Harold and of the Yale scholarly community; to Rocky Stinehour for reading my remarks on his early career and the circumstances of the merger and the consolidation; to Richard Benson for his help on the history and practice of photography, collotype, offset and related processes; to Nancy Hugo and Gregg Hugo for information on Harold's life, for access to documents bearing upon his work, and for their encouragement to undertake this project.

I am much obliged to Elizabeth W. Pope and Peg Lesinski of the American Antiquarian Society; to Lynn and Robert Veatch; to Virginia M. Adams (Mrs. Thomas R. Adams) for information on Harold's connections to the John Carter Brown Library, and to its then Director, Thomas R. Adams; to Dr. Neil Safer, Director, and Kim Nusco, Reference and Manuscript Librarian, of the John Carter Brown Library; to Eric Holzenberg of the Grolier Club; to Alexander Y. Goriansky, Keeper of the Archives of the Club of Odd Volumes; to Melissa Watterworth Batt of the Thomas J. Dodd Research Center, University of Connecticut, for assistance and access to the Rex Brasher Collection; to Morgan Swan, Special Collections Librarian, Dartmouth College Library; to Susan Odell

Walker and Kristen McDonald of the Lewis Walpole Library, Yale University; to Jae Rossman and Molly Dotson at the Robert B. Haas Family Arts Library, Yale University; to Elizabeth Frengel, Nancy F. Lyon and Anne Marie Menta of the Beinecke Rare Book & Manuscript Library, Yale University; and to Drika Purves, formerly of the Beinecke Library; to Suzy Taraba, Archivist and Special Collections Librarian, Wesleyan University; to Karen Bucky, Reference Librarian, Sterling and Francine Clark Art Institute Library; to Stephen R. Stinehour for illustration materials that have enhanced this publication; and to David Lorczak, my friend and former colleague at Meriden Gravure, for his help.

Janice Franco, now retired from the Meriden Public Library, has provided me with information and access to relevant materials in the Library's local history collection, as has Alan Weathers, Curator, and his staff at the Meriden Historical Society.

Photography has always been a big part of the technology and practice of printing at Meriden, and this has benefitted the illustrations in this volume. Robert Hennessey worked at Meriden, and he excelled in both reproductive photography and creative photography. Alan Rodgers, an account executive, was another with an enthusiasm and talent for photography. Their artistry is evident in those illustrations credited to them. B. A. "Tony" King and the late Carl Zahn, both devoted friends of Meriden, have made significant contributions to the visual record of the Meriden Gravure Co. I am much obliged to Tony and to his able assistant, Regina Zanetti, and to the family of Carl Zahn for the use of their photos.

I am very grateful for the help of Robert Hennessey. He has taken illustrations in a variety of printed and digital formats and has converted them to a single standard of the highest quality. In doing so he has also given me a valuable tutorial in some of the finer points of printing from digital copy. It has been my hope that the illustrations as printed would measure up to Harold's exacting standards. If they are considered so, it is due in no small part to Hennessey's interest, effort and expertise.

Finally, I am very much indebted to my publisher, Robert Fleck of Oak Knoll Press, for his support. This I take to be his expression of confidence that my book would be a worthy addition to the distinguished list of the Press. It is my great regret that he did not live to see this book to completion. It has also been my good fortune to work closely with Matthew Young, Managing Editor of Oak Knoll Press. I have profited much from his experience and good judgment, and I am appreciative of it.

Prologue

In 1960, having my college degree in hand and having satisfied my service requirement to Uncle Sam, I started out in search of a full-time job in my chosen field of printing and graphic arts. One of the companies that enjoyed an outstanding reputation nationally and internationally was a firm right in New England, right in my home state of Connecticut: The Meriden Gravure Co. I made an appointment for a job interview and on the agreed-upon day drove up to Meriden and found my way to Meriden's north end. I turned off Kensington Avenue and drove down Billard Street, looking for number 47. Billard Street was a short, grubby, dead-end street, lined with a few three-decker tenement houses, an auto body shop and a ramshackle collection of industrial buildings. Surely, I must have made a mistake, got the address wrong or taken a wrong turn somewhere. This could not possibly be the world-famous Meriden Gravure Company. But it was. The name on the door clearly said so.

I entered into a vestibule the size of a phone booth. An old lady with a steel-point pen perched behind her ear looked over the top of her rolltop desk and asked me what my business was. The place seemed to me like a scene out of a Dickens novel—the description of Tellson's Bank in *Tale of Two Cities* or Jacob Marley's counting house in *A Christmas Carol.* I told the lady that I was here to keep an appointment with Mr. Hugo. She admitted me and ushered me into the boss's book-lined office. The lady, I later learned, was Elsie Yale, the bookkeeper, and she took her duties very seriously. She was in charge of, among other things, office supplies. If anyone wanted a new pencil, they had to turn in to her the stub of the old one.

The interview with Mr. Hugo went well. My first impression of him was that he was an intelligent person, one very much committed to his work, the quality of his product and the satisfaction of his customers. He gave me a tour of the shop and explained that the company preferred to put its resources into the best equipment and machinery rather than a fancy building and expensive office furniture. The company had recently installed air-conditioning, but only out in the offset pressroom, where a steady and controlled temperature and humidity were necessary for the quality of the printing. The office employees, including all the top brass, could swelter. The Gravure had an opening for me, and other possibilities for employment were not so promising. And so it was that I began my twenty-five-year association with Harold Hugo.

FIG. 1.1. James Ferguson Allen as a college student and as a businessman in his early fifties. Both photos are from his Yale Class Report thirty years out of college. (Yale University Library, Manuscripts and Archives)

One

The Early Years of Meriden Gravure, 1888–1923

The Meriden Gravure Co. was founded by James Ferguson Allen, a young man born and raised in New Haven. After graduating from Yale in 1882, his family staked him to a start in life, which he used to purchase land in Montana in partnership with a Yale classmate. In doing so he was taking a path similar to that taken by his contemporaries, Theodore Roosevelt and Frederic Remington. All were Eastern college boys seeking their fortune in the Golden West. But their careers did not run parallel for long. Remington devoted himself to painting and sculpture in preference to ranching. Roosevelt was wiped out in the prairie blizzard of 1886 and returned to the East and plunged into politics. Allen, along with his partner, lost everything in the same blizzard, and he retreated home, a failure in the harsh judgment of his family.[1] Getting no sympathy from his own family, he turned to Charles Parker, the leading diversified industrialist of Meriden. Parker had been Meriden's first mayor when it adopted its city charter and was the occupant of the grandest mansion in town. Parker had backed two men, Frank Eaton and Carlton Peck, in the business of engraving halftone printing plates, but they did not work out to Parker's satisfaction. He forced them out and lent money to Allen to take over the business, the loan being on terms very favorable to the lender.[2]

At this time in the progress of printing technology, there were many people working on schemes to marry the new and rapidly improving process of photography to the mechanics of printing. There was photoengraving, photogravure, photolithography and various hybrid variants of these processes. All were attempts to produce illustrations quickly, accurately, economically and in large quantities. The most common illustration techniques at the time were wood engraving, copperplate engraving and stone lithography, all of which were slow, expensive handcraft processes. Among the processes being developed in Germany (where lithography was invented) was a process called "Lichtdruck" (literally "light printing") in German and photogelatin or collotype in English. (Collotype takes its name from the Greek word "kolla," meaning glue, and refers to the adhesive properties of the gelatin.[3] It should not be confused with Fox Talbot's Calotype process, which was a photographic process and not a photomechanical reproduction process.)

The twenty-eight-year-old Allen had no experience in printing or photography, but nevertheless saw the possibilities of collotype for commercial work. His plan was to do printing for the International Silver Company, Meriden's dominant industry.

The collotype process is considered a branch of lithography in the classification of printing processes because it is planographic, with the image neither raised above the surface of the plate nor incised into it. Further, it is lithographic because it works on the principle of the antipathy of greasy ink to water. The ink bears the image, and the water keeps the non-image areas free of ink. In a lithographic stone the moisture is held in the porousness of the stone. In a collotype plate the moisture is held in a gelatin skin that absorbs moisture. The gelatin is supported on a tempered glass plate about a half-inch thick. A light-sensitive potassium bichromate was mixed into the gelatin. This had the capacity to harden the gelatin and prevent the absorption of moisture in the gelatin in proportion to the amount of light that it was exposed to. This gelatin solution was flowed onto the plate, which had to be kept dead level so that the sensitivity of the coating remained uniform overall. The plate was heated to dry down the coating and then exposed to the continuous tone negatives arranged exactly as the images were to be printed to the sheet. The dry-down caused the gelatin to shrink a bit and take on a reticulated character. This reticulation, viewed under magnification, is a sure indication that the image was printed by collotype and not some other process. Following exposure, the plate was flushed out to remove the unexposed bichromate, the equivalent of fixing the film in the days of darkroom photography. When the plate was ready to go to press, the plate was soaked and the moisture was absorbed into the gelatin to the extent that it had not been hardened. This caused the gelatin to swell in the non-image areas and rise above the areas that were to take ink.[4] Although collotype was theoretically a planographic process, in actual fact it had some of the characteristics of an intaglio process. The recessed image areas held a depth of ink that transferred richly to the printed sheet. This transfer could only take place under heavy pressure; the weight of the press impression cylinder had to squeeze the paper slowly and laboriously into the plate on the bed of the press as it passed under the cylinder. A hundred years ago there were books and manuals that explained collotype platemaking and presswork in detail, but they never told the eager learner the whole story. Many practices critical to the success of the process varied from shop to shop and were passed on orally from the experienced to the inexperienced.

Allen became aware of the work that was being done in the New York studio of the highly successful landscape painter Albert Bierstadt. From the 1850s, when Bierstadt started doing his grand panoramas of the great West, he was interested in printing processes that could represent and promote his paintings. He had skilled wood engravers do small black-and-white versions of his paintings for *Harper's Weekly* and other popular magazines. Steel engraving was also employed. This was a more prestigious process, done in large sheet sizes and with much more artistic detail. Louis Prang did a chromolithograph of one of his important paintings in 1868. But chromolithography soon came to be associated with cheap and common work, and Bierstadt did not continue working with Prang.[5]

Bierstadt's older brother, Edward Bierstadt, was an experienced pho-

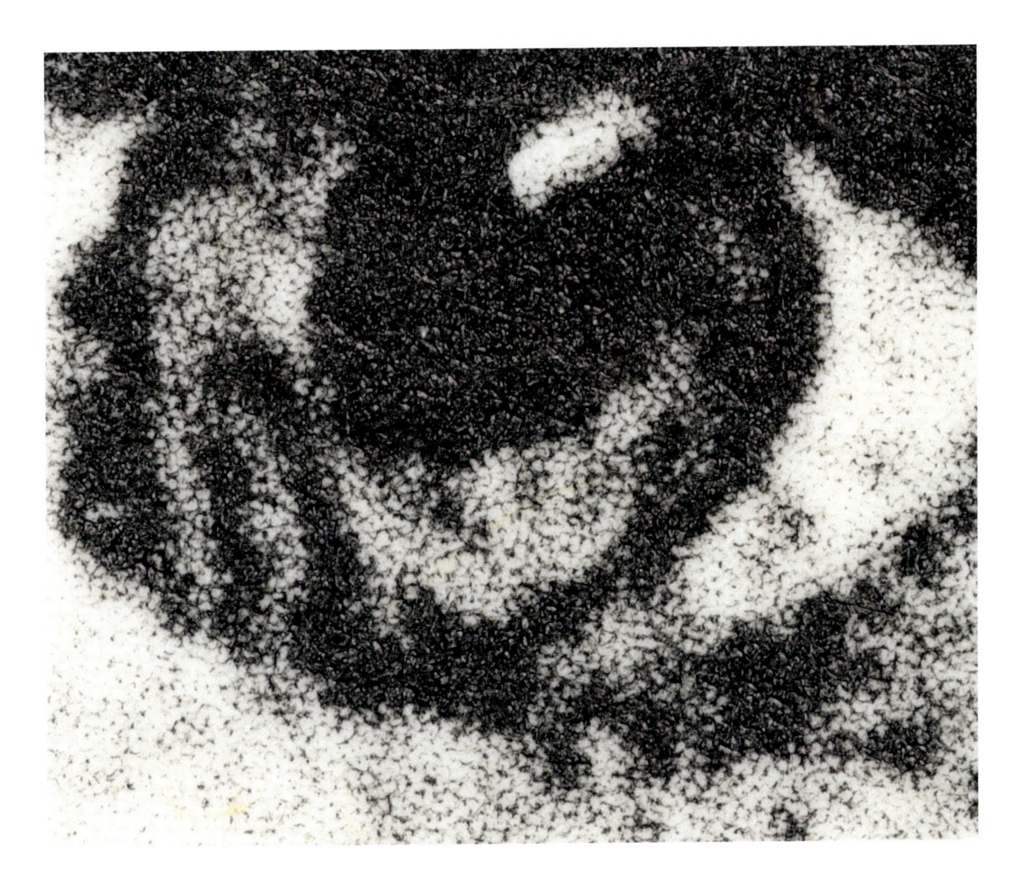

Fig. 1.2. Collotype enlargement to approximately twenty times actual size, showing the characteristic reticulation of the collotype grain. This is a detail of a collotype reproduction of a black-and-white Currier & Ives print.

tographer and photoengraver. He bought the American patents for the Albertype process from its inventor, Joseph Albert of Munich, in 1869. This variant of the collotype process greatly increased the plate life and made larger print runs possible, but at the expense of difficulties in exposure. His collotype work was of a very high quality, and in the 1880s he was in Albert Bierstadt's New York studio making collotype prints derived from his brother's paintings.[6] There seems to have been some ill feeling between Bierstadt

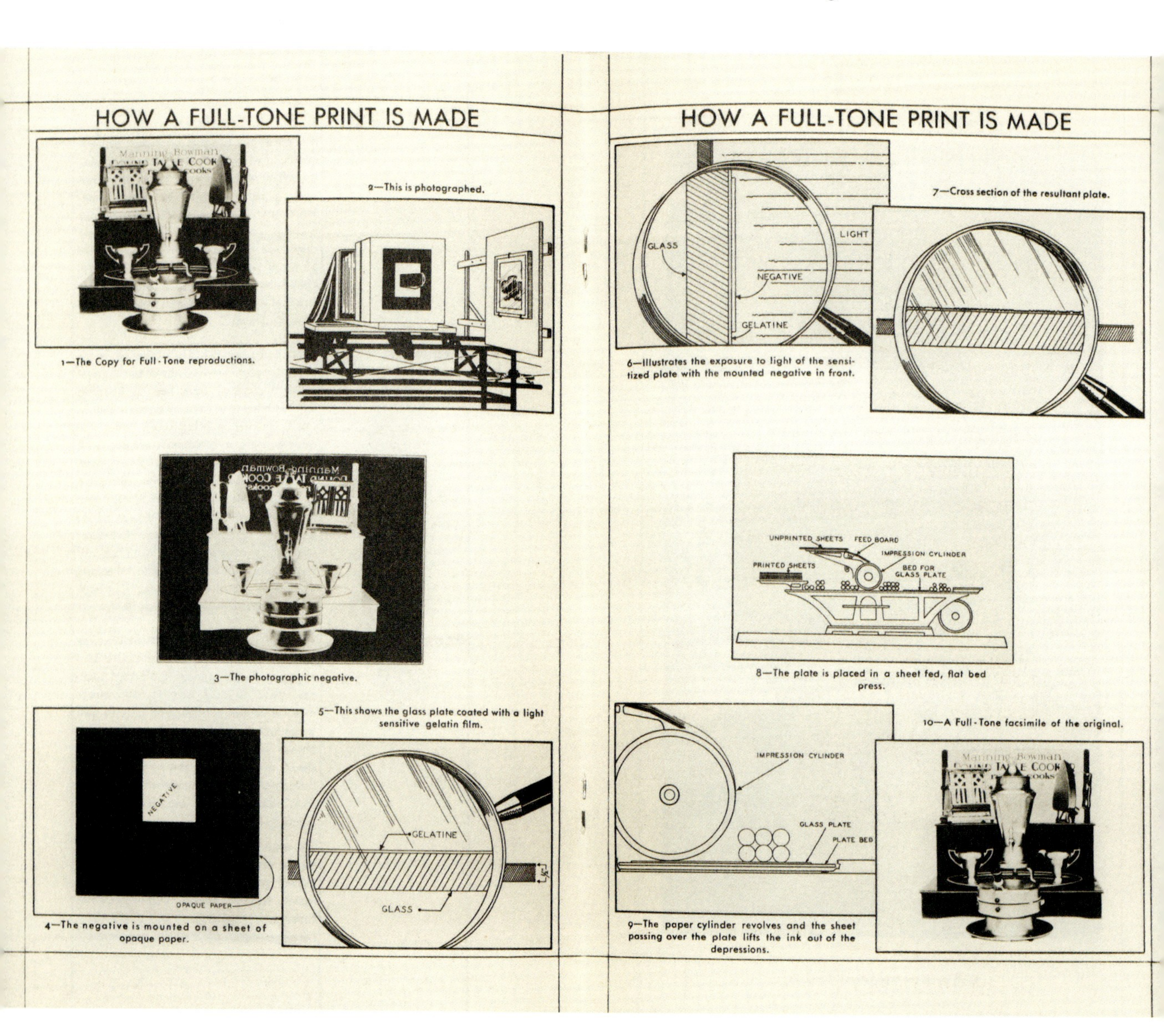

FIG. 1.3. A simplified explanation of the photo-mechanics of the collotype process. It makes no mention of the need to dampen the plate, which was done manually rather than mechanically. This is from a 1927 Meriden Gravure brochure promoting the process and its advantages to potential customers. (Meriden Historical Society)

and Allen because Bierstadt thought that Allen's commercial application of collotype degraded a process that ought to be reserved for fine arts printing. Nevertheless Allen persisted, and he imported German equipment and hired German craftsmen to run the equipment.

The Meriden Gravure Co. was formally incorporated in August 1888 with capital of $25,000 and with J. F. Allen as President and Treasurer. But it took some time to purchase equipment and organize production. Allen is listed as being employed by the Meriden Bronze Co. in 1889, by Eaton & Peck in 1890 and by the Meriden Gravure Co. as Treasurer in 1891.[7] In starting up the business, Allen was not without competition. There were a number of local photoengravers and printers. The large and well-established Forbes Lithographic Manufacturing Co. of Boston advertised the Albertype Process and actively sought the business of Meriden manufacturers.

The collotype output supported the national sales and promotion campaigns of the International Silver Co. Prior to this time, the sales force would call upon dealers throughout the country, showing wood engravings or actual samples of the silverware. Neither was satisfactory for the purpose. Wood engravings were stylized linear representations, which did not show the sculptural character of the silverware. Actual samples were bulky to carry and too valuable to leave with the dealers. With the improvement in print quality offered by the collotype process, printed samples supplanted these representations. This practice was so successful that other silver companies, competitors of International such as Towle, Gorham, Reed & Barton and Lunt, sought to have their printing done at the Gravure. The company prospered.

Collotype was not strictly speaking a gravure process, but it gave gravure-like results, and the name was more familiar and acceptable to American clients. It was for this reason that the corporate name The Meriden Gravure Co. came into being in 1888. In later years, the company proudly proclaimed its founding year on its letterhead, the date being easy to remember because it was the year of the great eastern blizzard. But that was not the blizzard that had ended Allen's ranching career.

In addition to its local work for International Silver, Meriden Gravure broadened its customer base to include other local manufacturers, industries with a national reach such as General Electric, Westinghouse Electric and Manufacturing Co., Western Electric and various glass manufacturers in the Pittsburgh area. Much of the work was printing loose-leaf illustration sheets to go in sales catalogues or parts catalogues.

In 1893 Allen married Cornelia Parker Breese, known familiarly as Nellie, granddaughter of Charles Parker.[8] One would have thought that in taking

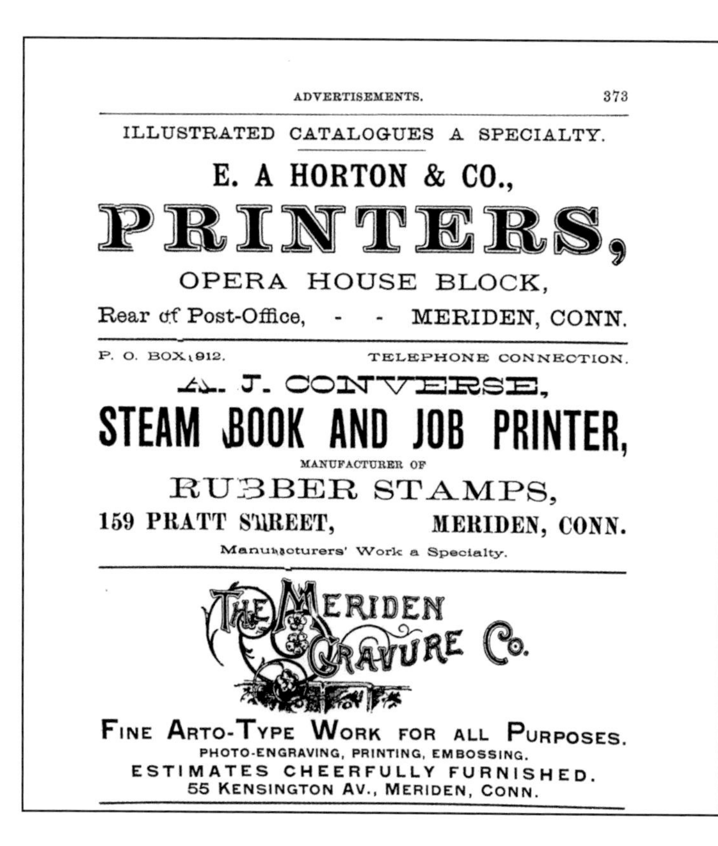

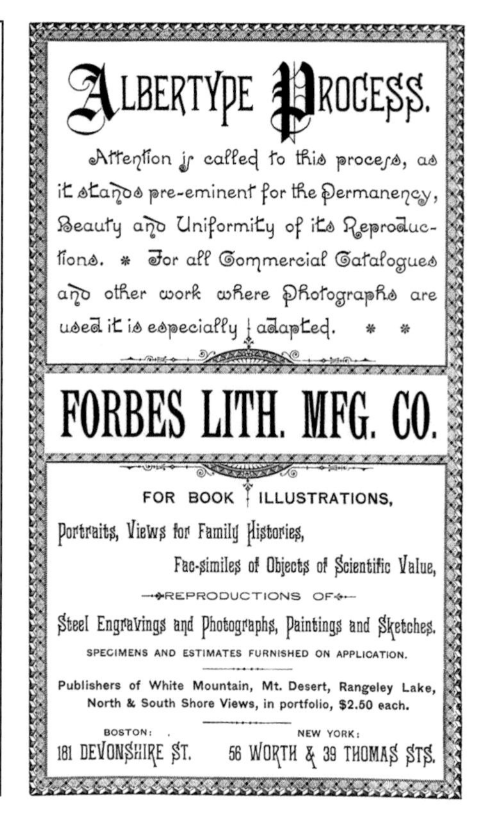

FIG. 1.4. Advertisement for the Meriden Gravure Co. in the best Art Nouveau style, along with ads for two other local printers. The property on which the plant was situated ran from Billard Street to Kensington Place, so either address could be used, but Kensington Avenue is an error. (1891 Meriden City Directory. Meriden Historical Society) FIG. 1.5. Advertisement for the Forbes Lithographic Manufacturing Co., featuring the Albertype Process. (1888 Meriden City Directory. Meriden Historical Society)

Allen into his family, Charles Parker might have forgiven Allen's debt to him, or at least relaxed its terms. But the original stringent terms seem to have remained in effect until the debt was fully satisfied in 1910.[9] Allen, for his part, was no less a taskmaster than Charles Parker. He was respected, but considered stern and somewhat remote. He paid his workmen, at least his pressmen, on a piecework basis, and only sheets that passed close inspection counted. But he also practiced small charities and was remembered by the kids in the neighborhood for giving out tickets to the circus when it came to town.[10] He was involved in local affairs as a board member of a commercial bank, a savings bank, the Boys Club and the Visiting Nurses Association.

Two years after the marriage, Nellie Allen gave birth to their first child, a son, Parker Breese Allen. Parker followed his father's path to Yale. He was

FIG. 1.6. The Meriden Gravure Co. plant, drawn for use in the 1906 *Centennial History of Meriden*. This was before the addition of the South wing. The glassed-in section of the North wing was used to make blueprint proofs by exposure to sunlight. The trees and shrubbery are an exercise in artistic license, as is the sign on the roof. The smoke belching from the chimney in those times was a sign of activity and economic prosperity, and not an indication of environmental degradation. (Meriden Historical Society)

a fine athlete but not a particularly talented or highly motivated student. His reputation was such that in later years there was the belief that he had been the first to score a touchdown in the Yale Bowl in its inaugural year, 1914. There is no truth to this,[11] but there is no question that he was an outstanding competitor both on the gridiron and in the swimming pool. In his younger days he went by the nickname of Babe. This may have been a flattering comparison to Babe Ruth. But it is more likely that he was being teased for his round, soft features that made him look baby-faced. In 1917 when war was declared against Germany, Parker interrupted his college career to enlist in the army and was posted to France as a non-commissioned officer in the artillery. Following the Armistice, he returned to New Haven. He received his bachelor's degree, *honoris causa*,[12] signifying that Yale recognized his service to the nation while excusing his uncompleted degree requirements. Parker went to work for the Dunlop Tire & Rubber Company in upstate New York. This was followed by a stint in New York City working for Bankers' Trust Company. In September 1923, J. F. Allen, now a widower living in rooms in the Winthrop Hotel, recalled his eldest son to Meriden to become his eventual successor. Parker just that summer had married Elizabeth Weeks, a Smith College graduate from Skaneateles, New York. Thus Parker, in his twenty-seventh year, was settling down both with respect to his career and with respect to his domestic life.

Parker's brother, Theodore Ferguson Allen, was two years younger. Ted Allen graduated from the Sheffield Scientific School at Yale and had a successful career as an engineer with IBM and Remington Rand. He never participated actively in the management of the Meriden Gravure Co.

Two

Harold Finds His Calling

Harold Hugo was born in Stamford, Connecticut (also the birthplace of Theodore Low DeVinne several generations earlier) August 8, 1910, to Otto and Esther Lundstrom Hugo, both of Swedish ancestry.[1] His full name was Everett Harold Hugo, but he disliked Everett and discouraged its use. In the workplace, he may have been called Everett behind his back, but never to his face. In more formal situations he was E. Harold Hugo and in more familiar ones, Harold. Moreover, he was Harold and not, if he could possibly help it, Harry or Hal or some other derivative nickname.

In 1917 the family, which now included two younger brothers, moved to the industrial city of Meriden. The family settled in a rental in the blue-collar north end of the city, a block from the Billard St. location of the Meriden Gravure Co.[2] Harold grew up an energetic and enterprising young man. At fourteen he got his working papers and applied to Parker Allen for a part-time job. He was unsuccessful in his initial application, but he was persistent and undiscouraged. He pestered Allen until he got the job and then set to work to make himself useful. His duties at this time were those of a general laborer: working in the shipping room, cleaning equipment and work areas, and transporting paper to and from the presses using the dumb waiter that passed for a freight elevator in the plant. He also helped in the chemical lab that was set up to research potential improvements in the collotype process, and recalled staying at the shop overnight on occasion to monitor the experiments.

Harold was a member of the Meriden High School Class of 1927. Most of his classmates would go directly to work in some local industry following high school graduation. Two of them would become long-term employees of the Meriden Gravure: Roswell Wuterich, who became supervisor of the collotype camera department, and Frank Bauchman, who ran the shipping and receiving department.[3]

Harold was among the minority with the academic ability and the motivation to continue his education. As a youth he was a good and active reader and particularly enjoyed adventure tales with illustrations by N. C. Wyeth.

Although he had strong leadership skills as an adult, he did not show these attributes in his high school career. He was not a class officer or a member of the Prom Committee, nor did he hold other similar responsibilities. He was not involved in sports, and his participation in extracurricular activities was sparse. He was, however, active in Scouting and earned enough merit badges to make Eagle Scout. Harold's family situation was very modest. His father held various jobs in local industry, but he would be characterized as a general laborer and not a skilled craftsman.[4] The second of the three Hugo brothers, Bert, came down with childhood polio. He eventually recovered from this but only after a period of care and therapy, the cost of which must have been difficult for the family. It is reasonable to believe that Harold curtailed his school activities to work part-time either to help the family out or to set aside money for college. Harold did a lot of bicycling in his youth. The Hugo family did not have a car, and this would help explain the bicycling. Harold

FIG. 2.1. Harold as a freshman member of the Northeastern University Class of 1931. (Hugo Family Papers)

learned to drive only after he went to work for the Gravure when Parker gave him keys to a car on which to learn.[5]

Harold matriculated at Northeastern University in Boston, the first member of his family to pursue a higher education. Harold did well in his freshman year in the School of Business Administration. His grades for the last of the four marking periods in the academic year were a B in English and in Finance and an A in Accounting, Business Administration and Economics.[6] However, in the fall of 1928, J. F. Allen died suddenly and Parker Allen succeeded him. Parker, who saw Harold's potential, persuaded him to join the firm as his full time assistant in the office. In his letter of withdrawal from Northeastern dated October 16, 1928,[7] Harold described his situation to the dean of the School of Business Administration as follows:

> I will not be able to return to school this fall for several reasons:
>
> First: I have accepted a position, estimating, with The Meriden Gravure Co. Mr. J. F. Allen, the president, died suddenly on Oct. 7, 1928. This left an opening that seems to be an opportunity. At the same time, this is a position which had to be filled immediately. Mr. P. B. Allen felt that I was qualified for the position, and offered it to me.
>
> Second: I am not in a position, financially, to successfully continue through the school year. I accepted the position with the reservation that if things do not turn out as well as expected, I can return to school next year. Then if I want to return, I will be better able financially to do so.
>
> A classmate of mine has consented to furnish me with the book list for the Sophomore B.A. course, and I intend to continue my studies at home.

Further formal education was not to be. It was well-intentioned but unrealistic to think that he could hold a responsible full-time job and continue his studies at home. Things for him turned out far better than he could have expected. At this point in his life he could not have imagined the career he would have with Meriden Gravure, or the esteem and affection that he would enjoy in the field of scholarly publishing. Unable to pursue to completion his course of study in business administration, he became something of an autodidact, studying informally to improve himself. But his choice of studies drifted away from the practice of business and into the world of books. Harold was both an attentive and retentive listener. Absorbing the circumstances and needs of his customers was an important and lasting part of his self-education. It was this interest and enthusiasm that endeared him to so many in the realm of scholarly printing and publishing.

Harold's responsibilities included estimating, but went well beyond that. He was active in sales and customer service and any other managerial work that Parker chose to delegate to him. Sales received a great deal of his attention in his first year of full-time employment. Part of his job was to send out sales letters with collotype samples targeting a wide variety of manufacturers including producers of glassware, lighting fixtures, home appliances, machine tools, post cards and greeting cards. This work from Meriden was supplemented by sales offices in Boston, New York, Philadelphia and Pittsburgh, and sales representatives in Atlanta, Rochester, San Francisco and Toledo.

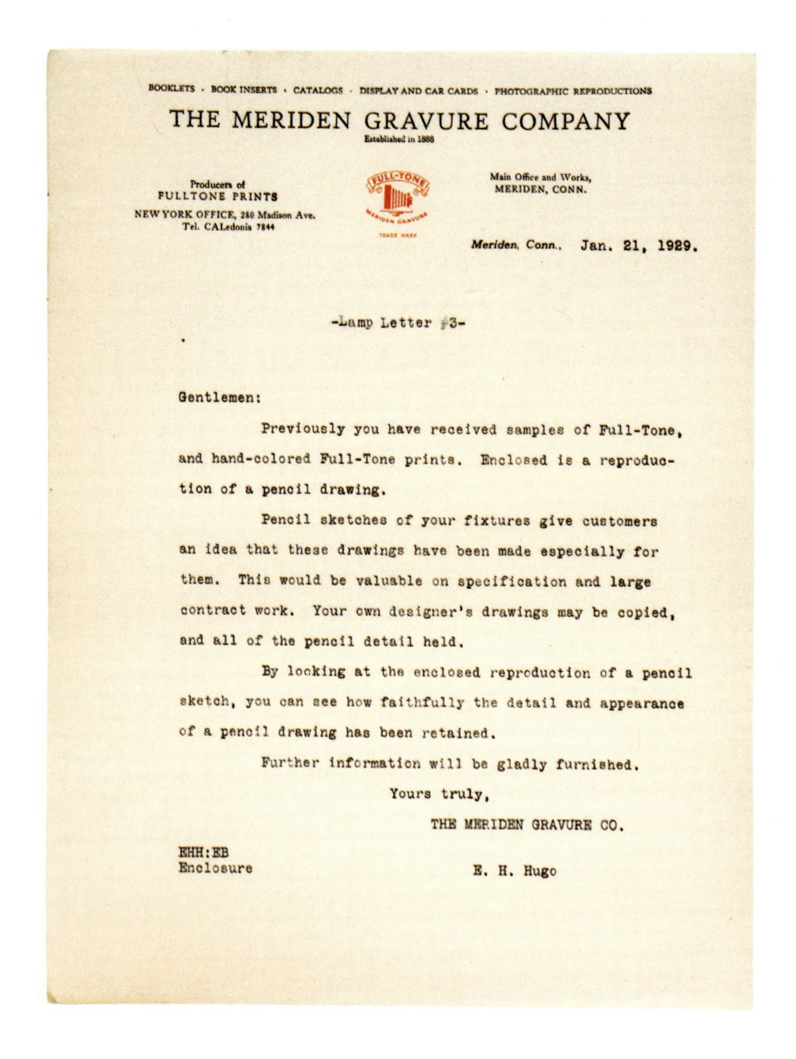

BOOKLETS · BOOK INSERTS · CATALOGS · DISPLAY AND CAR CARDS · PHOTOGRAPHIC REPRODUCTIONS

THE MERIDEN GRAVURE COMPANY
Established in 1888

Producers of
FULLTONE PRINTS
NEW YORK OFFICE, 280 Madison Ave.
Tel. CALedonia 7844

FULL-TONE
MERIDEN GRAVURE
TRADE MARK

Main Office and Works,
MERIDEN, CONN.

Meriden, Conn., Jan. 21, 1929.

-Lamp Letter #3-

Gentlemen:

Previously you have received samples of Full-Tone, and hand-colored Full-Tone prints. Enclosed is a reproduction of a pencil drawing.

Pencil sketches of your fixtures give customers an idea that these drawings have been made especially for them. This would be valuable on specification and large contract work. Your own designer's drawings may be copied, and all of the pencil detail held.

By looking at the enclosed reproduction of a pencil sketch, you can see how faithfully the detail and appearance of a pencil drawing has been retained.

Further information will be gladly furnished.

Yours truly,

THE MERIDEN GRAVURE CO.

EHH:EB
Enclosure

E. H. Hugo

FIG. 2.2. Three months after being hired full time, Harold was already active in the sales program. (Meriden Historical Society)

1930 was a good year for Meriden. They had a good backlog from the previous year, and the stock market crash had not yet reverberated throughout the economy as a whole. But 1931 was bad and 1932 was worse.[8] Business was off by 50% from what it had been. Universities and scientific organizations with conservatively managed endowments were much less affected by the bad economy than commercial ventures. Many had their collotype work done at prestigious European printers such as Emery Walker and Cotswold Collotype in England, Ganymede in Berlin and Jaffé in Vienna. A strenuous and successful effort was undertaken to capture some of this work. Meriden would compete with these vaunted European craftsmen.

As the Depression wore on, the sales offices were discontinued except for New York, and the use of sales representatives was abandoned. This meant that Harold had to do a lot of traveling to service far-flung accounts. Parker soon recognized that Harold had a talent for sales. He was not pushy, but he was energetic, resourceful and responsive to his customers and potential customers.

He was an optimist, but a tempered one. When he was calling on a customer in a congested urban area, his strategy was to drive directly to the address and expect to find a parking place there. Only if that proved unsuccessful would he settle for less convenient parking. It was an application of the golfing wisdom, "never up, never in." He frequently used the parking facility under the Boston Common after it was built. In taking his ticket he realized he was not getting a guarantee of parking, but rather, as he said good naturedly, "a license to hunt."

The Depression economy was hard going, but Harold, ever the persistent peddler, self-confident and undiscouraged by adversity, was successful in bringing back enough work to keep the Gravure in business and ultimately return it to profitability. Parker was anxious to diversify into scientific printing. This was publication work for universities, research organizations and scholarly societies. The Smithsonian Institution and the Carnegie Institution, both in Washington, and the American Museum of Natural History in New York were influential clients who brought the firm to the attention of other publishers, particularly in the geological sciences.[9]

In 1931 Meriden was given what was probably its first opportunity to work with the American Museum of Natural History. This was the printing of the plates for *The Permian of Mongolia* by A. W. Grabau.[10] This title was an extremely detailed scientific analysis of fossils collected on a series of expeditions taken by the AMNH, led by the intrepid Roy Chapman Andrews. Chester Reeds, editor of the publication, asked his friend and colleague, Carl O. Dunbar of the Peabody Museum, Yale University, to check out Meriden. Meriden's work was entirely satisfactory to Dunbar, and this led to a friendship and important working relationship between Harold and Dunbar. Grabau's work was reviewed in a geological journal,[11] which noted that the fossils are "excellently illustrated on thirty-five collotype plates."

This in turn led to a contract to print the plates for periodical and serial publications of the Geological Society of America and the American Society of Petroleum Geologists in Tulsa, Oklahoma. It also resulted in an introduction to Professor Ray Moore of the University of Kansas, a leading figure in the fields of geology and paleontology.[12] Among his many accomplishments, perhaps the greatest was the organizing and editing of the definitive *Treatise on Invertebrate Paleontology*, which was printed in many volumes over a period of years, beginning in the early fifties and continuing even after Moore's death in 1974.

In 1932, Arthur C. Sias, who is identified as Director of Research at the Meriden Gravure Co., wrote an article for the *Journal of the Biological Photographic Association*[13] in which he argued that while the text of scientific publications was usually accurately printed, the accompanying illustrations were often of poor quality. The value of a monograph was diminished by excessive retouching, distorted tonal values and loss of detail in both the highlights and shadows. The remedy for this problem was, first of all, to foreswear retouching and to use photographs that were accurate and detailed even at the expense of being pleasing. Then Sias got into his sales pitch. He explained the collotype process, remarked on its wide tonal range, its ability to hold detail and its economy in short-run publications. He concluded by noting that the illustrations accompanying his article had been produced by Full-Tone Collotype.

Allen also had ambitions of developing the fine arts market. He had established the "Full-Tone" brand name for his illustration work, and it remained in use for many years. Harold's duties included developing this market. The Full-Tone print program produced a variety of publications both large and small. One was a print reproducing artwork of a Connecticut historical landmark. Another was a depiction of a four-masted schooner proudly owned by an international yachtsman, work that came to Meriden through the agency of the well-known journalist-adventurer Lowell Thomas.

Later in the decade Meriden produced a series of promotional brochures, each one devoted to promoting collotype illustrations in a scientific discipline such as anthropology, archaeology, geology or paleontology. This paralleled the practice of targeting specific industrial products in earlier sales literature. In Supplement No. 5, devoted to Geology and Paleontology, Carl Dunbar echoed the arguments put forth by Sias: "The modern Full-Tone

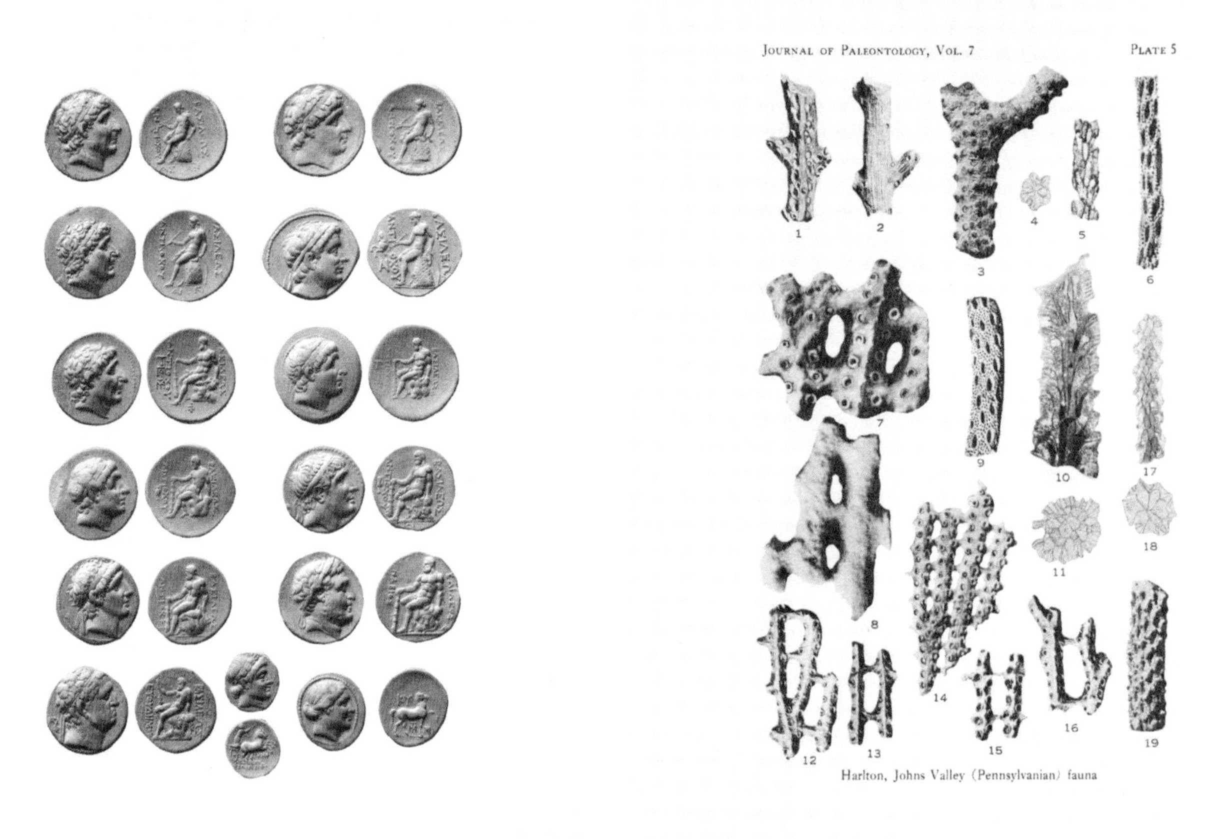

Fig. 2.3. A page of numismatic specimens dating from the early 1930s that originally appeared in a scholarly journal, and was overrun to use as a sample promoting Meriden's work. Fig. 2.4. A page of paleontological specimens that originally appeared in *The Journal of Paleontology*, Vol. 7, 1933. (Both illustrations Meriden Historical Society)

Collotype reproduction has provided us with a medium that can compete in price with the best halftone, and reproduces faithfully all details with the depth and fullness of a good photograph. Even faint details are faithfully shown . . . A special advantage of the Full-Tone Collotype reproduction is that the figures may be examined under a lens at magnifications up to ten diameters and more without being obscured by graininess which interferes in the case of half-tones." [14]

This program was expanded to include a brochure on the reproduction of works of art, written by Charles Rufus Morey of Princeton, a leading art historian of the time. Still later Philip Hofer, who was interested in books that were both scientifically important and attractive as art, contributed an illustration by a French horticultural engraver, appropriately named Bouquet, whose work was reproduced in collotype and hand colored by the Berrien Studio. This touted the practice of hand coloring collotype as an

FIG. 2.5. Hand-colored collotype illustration from the brochure *Full-Tone Collotype for Scientific Reproduction. Supplement No. 15, The Reproduction Series of the Harvard Library Department of Printing & Graphic Arts.* It dates from the 1930s.

alternative to printing color collotype, which Meriden was not able to master. The brochure program came to an end, but the involvement of such important figures as Dunbar, Morey and Hofer was helpful as an endorsement of Meriden's reputation for quality illustration printing.

One of the largest projects undertaken was done for the ornithological artist Rex Brasher of Kent, Connecticut. Brasher, a latter-day Audubon, made it his life's work to record all the birds of North America in their native habitats and produced 875 watercolor paintings, which were exceptional both for their scientific accuracy and their visual appeal. His magnum opus was a self-published set of twelve volumes issued in a limited edition of a hundred copies.

Birds and Trees of North America was originally to be produced at the prestigious and rather grandly named Printing House of William Edwin Rudge, of Mount Vernon, N.Y. Letterpress four-color process had been considered, but this was found to be prohibitively expensive. Rudge proposed using the Aquatone process, which used a conventional fine screen halftone exposed to a gelatin plate and run on an offset press. In Brasher's biography, *Rex Brasher, Painter of Birds*, the author, Brasher's nephew Milton E. Brasher, states that Brasher was dissatisfied with the Aquatone printing because it was too heavy and could not be hand colored successfully.[15] In 1929 Brasher turned to Meriden. Meriden submitted proofs of a sample watercolor in Full-Tone black-and-white collotype. Brasher was a severe critic of his own work and had destroyed many of his earlier paintings that did not measure up to the standards he was later able to achieve. He would be no less critical of the quality of the reproduction of his work. But Meriden achieved exactly the effects that Brasher was looking for. The black-and-white key plate proofs responded to hand coloring very well. Further, Harold and Parker were able to submit a better production schedule and more favorable financial terms than Rudge was willing to offer. The original paintings were sent to Meriden, and work was begun. As the printing progressed, the sheets were shipped in batches to Brasher, who did the critical hand-coloring of the birds personally and who closely supervised the background coloring. In addition to his artwork, Brasher provided information on the range of habitat and the nesting and breeding habits of each bird. He was a keen observer and was fascinated with the characteristic behavior of each species. His enthusiastic personal feelings regarding the Blue Jay reveal much of himself as well, and were expressed in part, as follows:

> One morning in late November when a gale was driving slants of rain thru the valley, I remembered a sledge I had left under a cliff in the woodlot. Glancing up at the squall-swept surface, I detected a fleck of blue. There were three Blue Jays snuggled in a niche thirty feet

Fig. 2.6. Rex Brasher, Blue Jay (female and male), Plate 477 from Volume 8, *Birds and Trees of North America*, (Publisher: Rex Brasher Associates, Kent, Connecticut; 1929-1932) considerably reduced. Note Brasher's monogram and the date of the painting on the stem of the oak leaf on the right. (Archives and Special Collections, Thomas J. Dodd Research Center, University of Connecticut, Storrs, CT)

> up, as comfortable as you please! There I left them—three rollicking buccaneers tamed by the elements so they did not even scold me. After all, isn't it those who, like the ancient Greeks, take life's storm and sunshine with a laugh, that we most admire?

Brasher was pleased with the result and appreciative of Meriden's effort. His biographer writes, "The compact with Meriden Gravure officers was altogether heartwarming and inspiring. Here were businessmen—almost total strangers, indifferent to security of any kind, willing and anxious to help a white-haired man realize a dream . . . It was indeed a tribute to Rex. It was a tribute to his work. There could be no other inference. Rex could not help feeling a great flooding surge of satisfaction and new confidence."[16] The officers referred to were Harold and Parker; the quotation recognizes the support and encouragement that Parker was prepared to give to this venture. Harold may have appeared to Brasher as being "indifferent to security," but he was not. One of Harold's great gifts was an instinctive sense of who could be trusted and who could not. He did not need to ask for bank references.

Enter Gregg Anderson

Gregg Anderson joined Meriden Gravure in 1932. He was a Californian who had worked at the Grabhorn Press and came east in the hope of getting a job with the Merrymount Press. Updike thought well of him but didn't have a position open. With Updike's recommendation he applied for a job with Richard Ellis of the Georgian Press in Westport, Connecticut. Ellis had no place for him either and referred him to Meriden. Meriden needed a compositor and took him on, which was a good thing because by this time Anderson was flat broke. Subsequently a vacancy arose, and he became head of the letterpress department. This included presswork as well as hand composition and makeup. Its principal function was to set and print captions for the collotype illustrations.

Anderson's predecessor as head of the letterpress department was Paul Johnston. Johnston in addition to being an experienced printer was also a bibliophile. He was the author of *Biblio-Typographica: A Survey of Contemporary Fine Printing Style*, printed by Fred Anthoensen at the Southworth Press and published by Covici-Friede in 1930. He was also the editor of *The Book Collector's Packet, A Monthly Review of Fine Books, Bibliography, Typography, & Kindred Literary Matters.* The first four issues were printed letterpress at Meriden Gravure and issued from Meriden. Johnston left Meriden in the summer of 1932, and subsequent issues were published from Woodstock, New York, where Johnston had set up his own shop. Harold might have been expected to be attracted to Johnston and his bibliographic interests, but this seems not to have been the case. In announcing his departure for Woodstock, Johnston wrote, "Woodstock, it seems, with its pleasant surroundings and associations will be a better place for producing *The Packet* than Meriden was." This acerbic remark suggests that Johnston may not have left Meriden on the best of terms with Parker and Harold.

If Johnston was not congenial to Harold, Anderson was. Both were bachelors at the time. Harold tried to interest Anderson in joining him evenings at the speakeasies. Anderson was one of the dedicated, one who labored in the trade by day to make a living, and printed at night on his own time for the satisfaction of doing so. He talked Harold out of his nocturnal habits and introduced him to the dangerous addiction of printing for pleasure.[17] It was Anderson who taught him the rudiments of typography and letterpress printing.

The first fruit of this addiction was *Some Notes on Early Connecticut Printing* by Albert Carlos Bates, Librarian of the Connecticut Historical Society. Harold had previously done a few small jobs for Bates, so they were known to one another. The text had previously been published in the *Papers of the Bibliographical Society of America.* Harold proposed that the text be set anew and issued as a separate publication. Bates agreed to this, as did the editor of the *Papers* and the University of Chicago Press, publisher and distributor of the *Papers.* All of this was on the understanding that the publication was to be a limited edition, and not for commercial sale. There was a considerable interest in local American printing history at this time. Edmund Thompson at Hawthorn House in Windham, Connecticut, produced some charming pieces on the subject. The Book Club of California, with which Anderson was familiar, was active in this area. The choice of Bates's piece

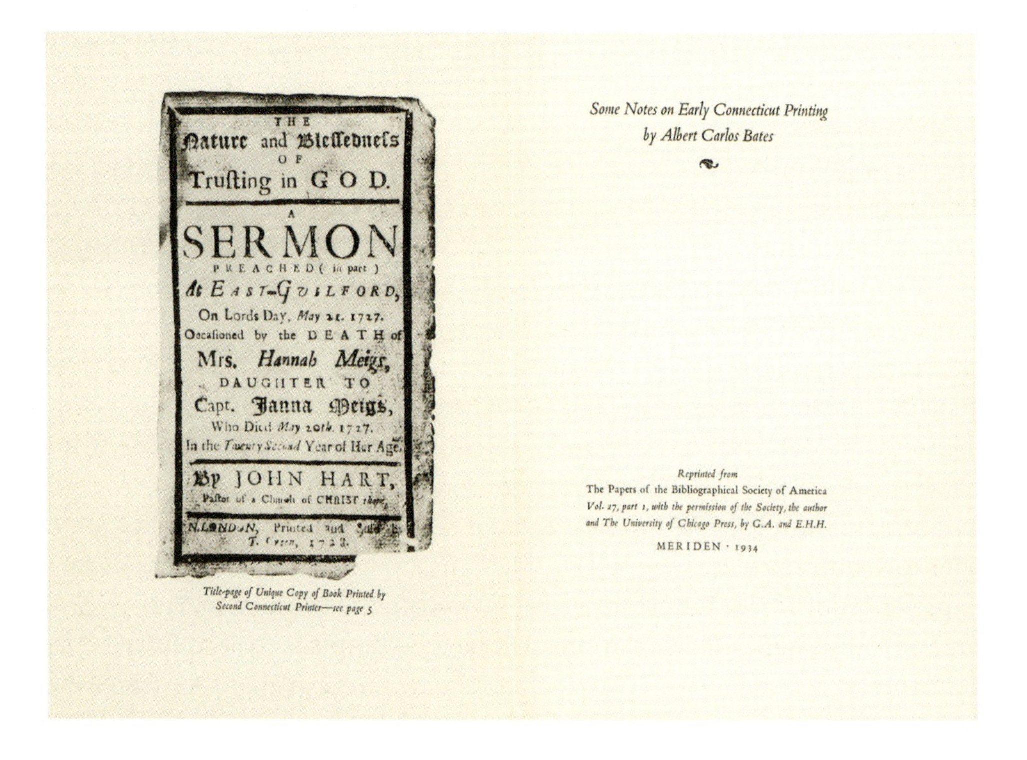

THE
Nature and Blessedness
OF
Trusting in GOD.

A
SERMON
PREACHED (in part)
At EAST-GUILFORD,
On Lords Day, *May* 21. 1727.
Occasioned by the DEATH of
Mrs. *Hannah Meigs*,
DAUGHTER TO
Capt. Janna Meigs,
Who Died *May* 20*th*. 1727.
In the *Twenty Second* Year of Her Age.

By JOHN HART,
Pastor of a Church of CHRIST *there*,

N.LONDON, Printed and

Title-page of Unique Copy of Book Printed by Second Connecticut Printer—see page 5

Some Notes on Early Connecticut Printing
by Albert Carlos Bates

Reprinted from
The Papers of the Bibliographical Society of America
Vol. 27, part 1, with the permission of the Society, the author and The University of Chicago Press, by G.A. and E.H.H.

MERIDEN · 1934

FIG. 2.7. Frontispiece and title page of Albert Carlos Bates, *Early Connecticut Printing*, as reprinted by Gregg Anderson and Harold Hugo.

therefore was logical. Harold managed the negotiations with outside parties, did the composition and handled the distribution of the finished product. Anderson was responsible for the design, proofreading and presswork. The planning of the publication began in the summer of 1933, and finished copies were ready for distribution the following spring.

Early Connecticut Printing made up to twenty-four pages including front matter. It was printed letterpress with a collotype frontispiece illustration. Printed on odd lots of paper, it took eight different lots to produce the eighty-three finished copies.[18] The publication received favorable notice in several publications, despite the fact that it was not commercially available and any public demand for it could not be satisfied. Edmund Gress gave it brief but approving mention in *The American Printer.*[19] Carl Rollins called it charming and gave it notice in his column, "The Compleat Collector," in *The Saturday Review of Literature.*[20]

The distribution of *Early Connecticut Printing* brought Harold to the attention of a number of people who would become personal friends and important customers: Clarence Brigham, Director of the American Antiquarian Society, Worcester; Randolph Adams, Director of the William L. Clements Library at the University of Michigan; Lloyd A. Brown, Curator of Maps at the Clements Library; S. T. Farquhar, Manager of the University of California Press, Berkeley; Melvin Loos of the Columbia University Press; John M. K. Davis, of Case, Lockwood & Brainard, printers of Hartford; David Pottinger, Associate Director of the Harvard University Press; Roland Baughman, of the Huntington Library; and Helmut Lehmann-Haupt, Curator of Rare Books at the Columbia University Library. [21]

Anderson resigned from the Gravure in 1935 because his wife had tuberculosis that did not respond to treatment in the East. She was advised that the chances for recovery would be better back in California. Anderson moved back to the Los Angeles area and joined the Ward Ritchie Press and subsequently formed the partnership of Anderson & Ritchie.

Even a continent away, Anderson was a presence and a lasting influence in Meriden. Harold was profoundly grateful to Anderson for introducing him to the practice of typesetting and letterpress printing, a life-changing experience for Harold. In a letter written in 1970 when Harold had received an honorary degree from Wesleyan University, he said to Ward Ritchie, "To return to the subject of honorary degrees, I can't help but feel that mine at least was greatly, or I should say tremendously influenced by my early years with Gregg Anderson because as I look back, if he had not spent those years in Meriden, I think the Meriden Gravure Company would be a very different

FIG. 2.8. Gregg Anderson with Caroline Anderson, his second wife, in 1944 shortly before being shipped overseas. From *Gregg Anderson at the Meriden Gravure Company*, 1946.

place today. As it worked out, we have not made a lot of money but we seem to have accomplished something that to some of us is more worthwhile."[22]

The typefaces that Anderson bought for the letterpress department were used by his successors (but with less skill) until that department was discontinued in the 1960s. (see Fig. 3.19) The typographic refinement of the folders that contained the collotype samples was probably lost on the scientists to whom they were directed. But Anderson's use of ligatures, swelled rules and drop cap Weiss initials gave these publications a typographical distinction in addition to promoting the quality of reproduction.

It is likely that Anderson had some ongoing connection with Meriden as its West Coast representative and was either on a retainer or on commission. It is certain that he was included on the Meriden Gravure Company Honor Roll listing those employees who were called to the colors during World War II.[23]

The southern California climate did not prove to be an effective therapy for Bertha Anderson, and she died in June 1937. When World War II broke out, Anderson volunteered for military service and was commissioned as a second lieutenant in the infantry. He was killed in action in Normandy in July 1944. Harold's only son, born the following year, was named Gregg in memory of Anderson. Harold was moved to write and have printed a brief but appreciative tribute to his devoted friend. *Gregg Anderson at the Meriden Gravure Company* appeared in 1946 in an edition of 350. Harold did not write much for publication, and it is noteworthy that he chose not to have his name as author on the title page, although he did sign the article at its conclusion. *To Remember Gregg Anderson: Tributes by Members of the Columbiad Club, The Rounce and Coffin Club, The Roxburghe Club and the Zamorano Club* was printed in 1949 for private circulation. This publication included a reprinting of Harold's piece, contributions from Gregg's California friends on his years before and after Meriden, an insert of illustrations printed in collotype at Meriden and Lawrence Clark Powell's bibliography of his printed work.

The Timothy Press

Encouraged by the success of *Early Connecticut Printing*, the next after-hours project undertaken was Anderson's informal memoir of his time working with the Grabhorn brothers in San Francisco, *Recollections of the Grabhorn Press*. This was published in 1935 in an edition of seventy copies and became the first of many Columbiad Club keepsakes. As with *Early Connecticut Printing*, most of the copies of *Recollections* went to friends of Harold and Gregg and to customers or prospective customers of Meriden Gravure.

The DeVinne & Marion Presses: A Chapter from the Autobiography of Frank E. Hopkins followed. This was a considerably more ambitious publication of seventy-two pages in an edition of three hundred fifteen numbered copies, of which two hundred and fifty were for sale. *The DeVinne & Marion Presses* was designed by Anderson in Los Angeles and produced by Harold in Meriden. The title page bears the date 1936, but it was not bound and ready for distribution until well into 1937. This was the first of Harold's publications to carry his adopted press name, The Timothy Press, although he retroactively considered his previous two efforts to have been printed by the then unnamed press. The Timothy Press took its inspiration from a dynasty of Connecticut printers, all named Timothy Green. As explained in *Early Connecticut Printing*, Timothy Green was the second official state printer and served in this office from 1714 to

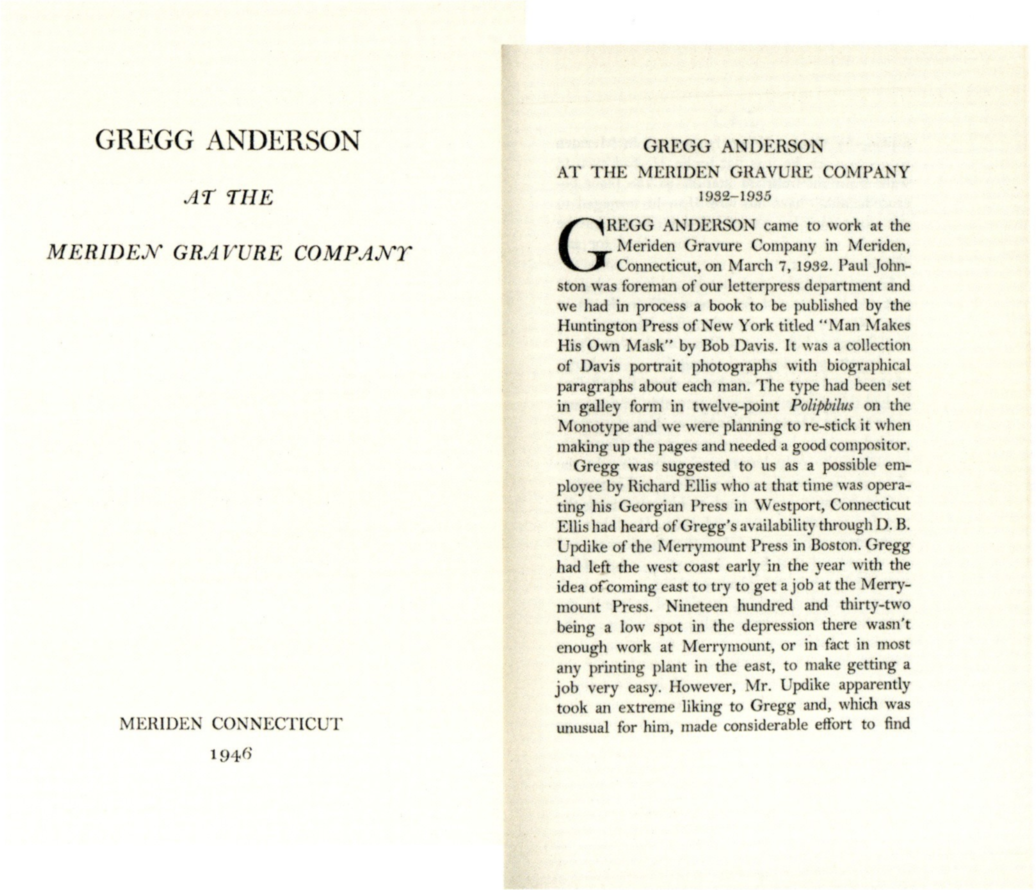

GREGG ANDERSON

AT THE

MERIDEN GRAVURE COMPANY

MERIDEN CONNECTICUT

1946

GREGG ANDERSON

AT THE MERIDEN GRAVURE COMPANY

1932–1935

GREGG ANDERSON came to work at the Meriden Gravure Company in Meriden, Connecticut, on March 7, 1932. Paul Johnston was foreman of our letterpress department and we had in process a book to be published by the Huntington Press of New York titled "Man Makes His Own Mask" by Bob Davis. It was a collection of Davis portrait photographs with biographical paragraphs about each man. The type had been set in galley form in twelve-point *Poliphilus* on the Monotype and we were planning to re-stick it when making up the pages and needed a good compositor.

Gregg was suggested to us as a possible employee by Richard Ellis who at that time was operating his Georgian Press in Westport, Connecticut Ellis had heard of Gregg's availability through D. B. Updike of the Merrymount Press in Boston. Gregg had left the west coast early in the year with the idea of coming east to try to get a job at the Merrymount Press. Nineteen hundred and thirty-two being a low spot in the depression there wasn't enough work at Merrymount, or in fact in most any printing plant in the east, to make getting a job very easy. However, Mr. Updike apparently took an extreme liking to Gregg and, which was unusual for him, made considerable effort to find

FIG. 2.9. *Gregg Anderson at the Meriden Gravure Company,* Harold's tribute to his dear friend. Designed by Carl Rollins and printed at the Yale Printing Office.

1756. The fourth state printer was another Timothy Green, the son of the earlier Timothy Green, and the fifth state printer was the nephew of the fourth state printer and also named Timothy Green.

Later Timothy Press productions were "Come Over and Stay 'Til Doomsday," a chapter from *The Cloister and The Hearth* by Charles Reade, ten printed pages with decorations by Valenti Angelo in his characteristic Italianate style. One hundred copies were issued under Angelo's Golden Cross Press imprint in 1937. The following year saw the publication of *Loyalist Operations at New Haven,* an account of the 1781 Loyalist raid on New Haven, edited by Lloyd A. Brown of the Clements Library, based on documents in the Clements collection. The fifty-three copies printed were divided between

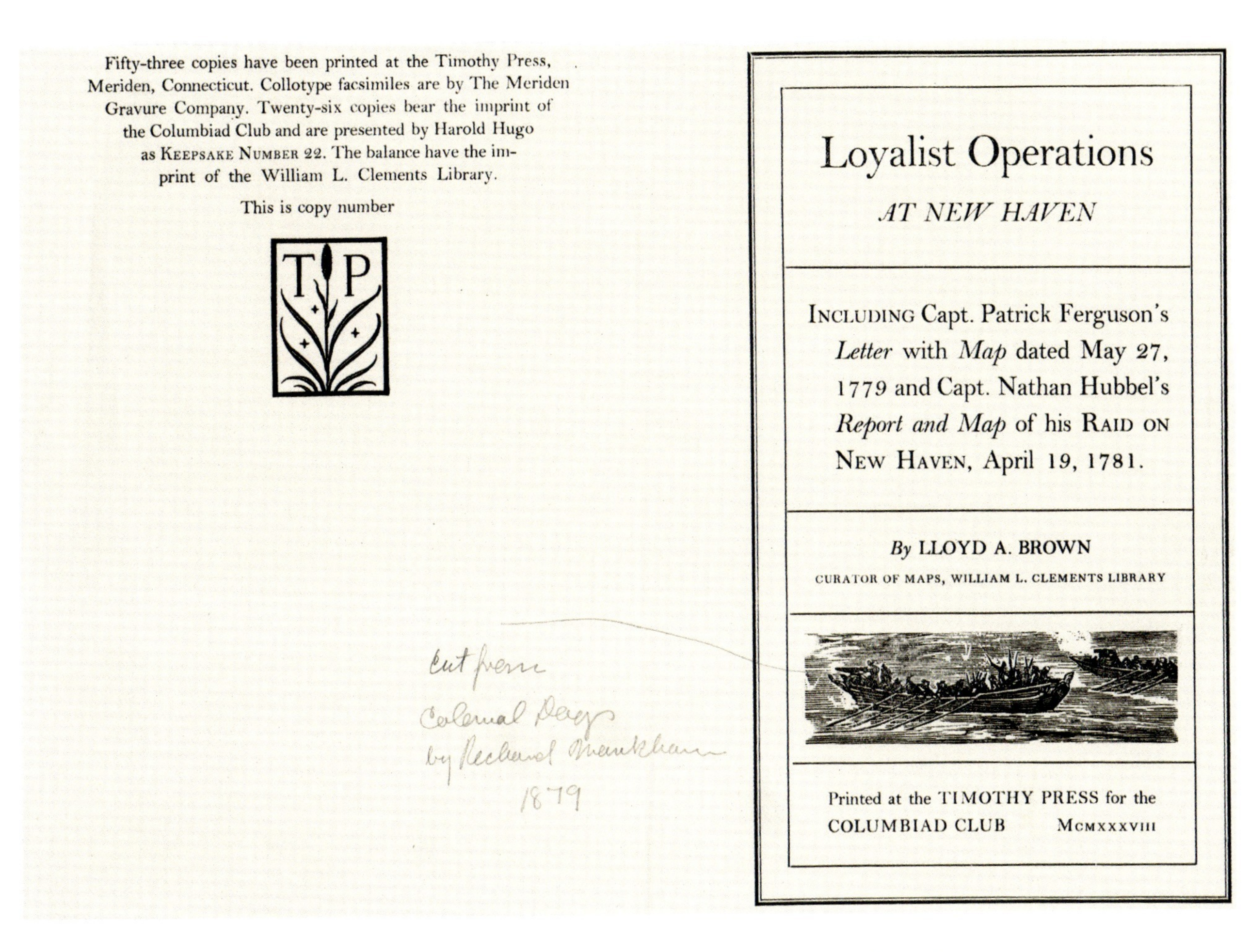

Fifty-three copies have been printed at the Timothy Press, Meriden, Connecticut. Collotype facsimiles are by The Meriden Gravure Company. Twenty-six copies bear the imprint of the Columbiad Club and are presented by Harold Hugo as KEEPSAKE NUMBER 22. The balance have the imprint of the William L. Clements Library.

This is copy number

T P

Cut from Colonial Days by Richard Markham 1879

Loyalist Operations

AT NEW HAVEN

INCLUDING Capt. Patrick Ferguson's *Letter* with *Map* dated May 27, 1779 and Capt. Nathan Hubbel's *Report and Map* of his RAID ON NEW HAVEN, April 19, 1781.

By LLOYD A. BROWN

CURATOR OF MAPS, WILLIAM L. CLEMENTS LIBRARY

Printed at the TIMOTHY PRESS for the COLUMBIAD CLUB MCMXXXVIII

FIG. 2.10. Proofs of the colophon and title page of *Loyalist Operations*. Pencil notations are in Harold's hand. The pressmark of the Timothy Press was designed by Valenti Angelo. (Hugo Family Papers)

the Clements Library and the Columbiad Club, with the latter copies being designated as Keepsake 22.

In 1941 Harold produced the first and only Timothy Press imprint that was primarily illustrations rather than text. This was *A Facsimile of Gribelin's Book of Ornaments*, reproducing in collotype six engraved plates from the collection of Philip Hofer of designs by the baroque engraver Simon Gribelin. Hofer, who suggested the publication, wrote a brief biographical introduction and appreciation of the craftsman. This was followed by a technical commentary on the reproductive process by Rudolph Ruzicka.

Several small pieces intended as Columbiad Club keepsakes appeared in 1941 and early 1942. The war curtailed the activities of the Timothy Press.

In the postwar period the Press was resumed, but not with its former vigor. *Publisher's Dilemma* was a four-page piece of ephemera with the Timothy Press imprint issued in 1948 to mark Herbert Bittner's fiftieth birthday. Harold co-published *Abel Buell, A Jack of All Trades & A Genius Extraordinary* with Tom Harlow, Director of the Connecticut Historical Society and longtime Clerk of the Columbiad Club. This publication is the last Timothy Press imprint known to the author. Harold distributed it to the members of the Columbiad Club as Keepsake 56 in 1955.

The Columbiad Club

The following is Harold's account of the founding of the Columbiad Club, as it appears in *Gregg Anderson at the Meriden Gravure Company*.

> I went to the west coast in the fall of 1934 as I had for two or three years previously, and came back much impressed with the accomplishments of the California book clubs, the Book Club of California, the Roxborough Club of San Francisco, the Zamorano and the Rounce & Coffin Club of Los Angeles. Gregg and I got to talking over the possibility of establishing something similar in Connecticut. In December 1934 we decided to do something about it. One Saturday afternoon we drove over to Windham, Connecticut to visit Ned Thompson. We had heard of him and knew his work, but had never met him. We spent a most delightful afternoon at Hawthorne House. Ned was enthusiastic about forming a printers book club. We then journeyed down to New Haven one afternoon to call on Carl Rollins of the Yale University Press. The idea met with his enthusiastic approval as it was something he had had in mind for a long time.

A dozen other bibliophiles were quickly recruited, and regular meetings commenced early the following year. Additional members were added, but the membership was never intended to be more than fifteen or twenty, the number that could be comfortably seated around a large, convivial dinner table. There were no constitution or by-laws, no Robert's Rules, and no election of officers. Each meeting was presided over by a moderator who was responsible for the evening's program, and this position was rotated through the membership. The only permanent positions were the Clerk (Secretary) and the Bailiff (Treasurer), which were positions held for life, or at least until retirement or resignation. There was never any ambition to have a clubhouse or a club library. This intimate informality was motivated largely by

Harold's dislike of bureaucracy and by Rollins's socialist principles. In 2010, the Club celebrated its seventy-fifth anniversary under the same minimalist governance established by its wise founders.

The name Columbiad Club is believed to have been the idea of Gregg Anderson. *The Columbiad* was a narrative poem written by Joel Barlow, one of the Hartford Wits. It was a verse epic on the founding of America, the intended worthy equal of *The Aeneid,* Virgil's celebration of the founding of Rome. To the Connecticut bibliophiles, its total lack of literary merit was redeemed by the excellence of the typography and printing, done in the best style by Fry & Kammerer of Philadelphia in 1807.[24]

Fig. 3.1. Parker Allen in uniform, 1942. This was produced in collotype in an edition of several hundred copies for the J. T. White Co., which solicited business and professional men to have their portraits printed for distribution to family, friends and associates. (Meriden Historical Society)

Three

Meriden Reinvents Itself

During the late twenties and into the thirties a great deal of research was devoted to improving the quality, consistency and productivity of collotype. Many in the industry believed that the fruits of research should be shared by all for the general good of the industry. As one of the technological leaders however, Meriden felt that it had more to lose than gain by such a policy. It was not inclined to share its hard-earned knowledge with competitors. Much of the research effort at this time focused on attempts to produce collotype on rotary presses with automatic feed and delivery and mechanically precise and consistent sheet registration systems. A number of outside consultants were employed to make recommendations on technical matters. Collotype and offset lithography were both basically chemical processes. Chemists with expertise in colloids, suspensions and photosensitive agents were hired to advise.

In 1926 a patent application was made by Karl Davis Robinson of New York for a "method of electroplating on metal articles which consists in coating the metal with gelatin prior to the electroplating step, drying the gelatin, causing the gelatin to swell to the required condition of permeability, then plating on the articles a metal through the gelatin in an electrolytic bath, and drying the gelatin coating." This apparently was an attempt to create a more durable plate than the gelatin plates then in use. It was never adopted as a production process at Meriden either because it was impractical or because the quality of the resulting printed image was not satisfactory. The patent was assigned to The Meriden Gravure Co., which suggests that Robinson was working for Meriden on this technique.[1] It was quite probably this work that occupied Harold when he had night duty in the research lab during his high school years.

Michael Bruno was on the Meriden Gravure payroll in the early thirties. He was a New Haven native and 1931 graduate of Yale's Sheffield Scientific School, where he excelled in chemistry and physics.[2] He became interested in collotype and was hired by Meriden no doubt to apply his scientific training to the improvement of the process. His length of service to Meriden and the nature of it is unrecorded. However, it is known that by 1937 he was married and living and working back in New Haven. But the Meriden employment was the small beginning of his lifelong dedication to the advancement of printing technology,[3] as author, teacher, Director of the Lithographic Technical Foundation (now the Graphic Arts Technical Foundation, GATF) and as Manager of Graphic Arts Research for International Paper Co.

Clarence Kennedy became a consultant to Meriden in the 1930s, and his consultancy continued until his death in 1972.[4] Kennedy was a professor of art history at Smith College who learned to be an excellent photographer of sculpture and of the dark interiors of churches in the course of his researches. He was employed by the well-known art dealer Joseph Duveen to photograph Duveen's best and most expensive sculpture for sale on the art market. Kennedy's great interest was in capturing the three-dimensional qualities of sculpture in all their nuance and subtlety. He desired to improve the quality of both printed reproductions and of lantern slides used in art history lectures. Silverware was essentially sculptural, and the Gravure had from its very beginnings wrestled with the problem of its effective reproduction. In the mid-thirties, its work for the International Silver Company had fallen off, but Meriden was trying to position itself in the art reproduction market. The work that Kennedy was doing was of great interest to Parker and Harold. Most of his contributions were in the early years of this association. Kennedy greatly admired the depth and richness of Meriden's black-and-white collotype. In 1940 he commissioned Meriden to print in collotype sets illustrations of important works of art for use in the Smith College survey course in Western Art. In 1948 Oxford University Press published *The Renaissance Painter's Garden* by Kennedy's wife, Ruth Wedgwood Kennedy, with design and hand-typesetting by Clarence Kennedy and collotype illustrations printed at Meriden. It was one of the thirteen books described by John Peckham in *Adventures in Printing.* In 1962, Leonard Baskin's Gehenna Press issued *Four Portrait Busts of Francesco Laurana* by Ruth Wedgwood Kennedy, illustrated with twelve photographs by Clarence Kennedy. These were reproduced in offset by Meriden, and they attest to the advancement and acceptance of offset reproduction by that time.

Kennedy in the thirties was also working with Edwin Land. Both were much interested in advancing the use of three-dimensional, stereoscopic optical techniques in the practice of photography. Land, through Kennedy, also became a consultant for Meriden. It is not clear what Land's contribution to Meriden might have been. He held many patents, but there is no indication that any were licensed to Meriden. It is more likely that his advice was general; he was familiar with current work in optics and in photosensitive chemicals and had an idea of which lines of research were likely to be most fruitful.

Meriden was not successful in the effort to develop rotary collotype largely because Meriden's characteristic rich collotype effect was achieved by a heavy lay of ink that would not dry sufficiently quickly under high-speed conditions. Meriden's collotype work was always done on slow, flatbed reciprocating presses. Flysheets had to be inserted between each printed sheet to prevent the fresh ink from getting on the back of the previous sheet. Both the press sheets and the flysheets were fed by hand. A collotype press crew consisted of a team of three: the master pressman, his feeder-helper and the flysheeter. These presses had no automatic inking or dampening systems. When the plate needed more ink or more moisture, the press was stopped and the rollers were replenished with ink or the plate was replenished with moisture, both done manually. As a result, sheets from the run were never 100% consistent. When a customer would point this out, Parker's upbeat rejoinder was "No two alike, but all good."

All of the problems of printing collotype in black and white were compounded exponentially when it came to trying to print color. The inconsistency of the inking and dampening on multiple passes through the press led to many more sheets failing inspection. With each pass through the press, the paper absorbed some moisture from the plate and expanded, making exact register across the sheet impossible. Further, the sheets could not be hand fed into the cylinder grippers multiple times with sufficient consistency. The result was much greater spoilage allowances, more press time and more expense for paper and ink.

Two notable attempts to overcome these obstacles were undertaken. The first was the color frontispiece for Sachs and Mongan, *Drawings in the Fogg Art Museum*, published in 1938 in three-color collotype. It was an artistic success, and Agnes Mongan, the co-author, became Harold's friend for life. But the cost overruns were significant, and the job had to be rationalized as an expensive learning experience.[5]

The second publication was the first volume of *The Ledoux Collection of Japanese Prints of the Primitive Period.* It was done in 1941 for the Japan Society and had twenty color plates. That was also a success in the eyes of the customer, but as Harold said, "we used as many as twenty-five colors. That cost us our shirts, too." [6]

As a result, Meriden in the 1930s and afterwards yielded much of the color collotype printing to Jaffé. As late as 1962 the printing of the color collotypes for the *Chinese Calligraphy and Paintings in the Collection of John M. Crawford, Jr.* went to Jaffé in New York.

The technical expertise and offset press equipment brought in for these experimental purposes in the late thirties were to start Meriden down the path to high-quality photo-offset printing in subsequent years. Harold and Parker worked well together and agreed on most things. Parker was becoming frustrated and impatient with collotype, and he was the stronger proponent of developing offset. Harold felt that the firm was so strongly identified with collotype and the Full-Tone brand associated with it, that collotype could not be abandoned. He was inclined to continue to try to overcome the deficiencies of collotype by further research and experimentation. They compromised and pursued both until the nation went to war. The war forced changes on them, and the changes tipped the scales in the direction of offset development.

Harold's rise in the company was accelerated by the outbreak of World War II. Parker, who had served in World War I, volunteered for Army service in 1942 and was commissioned a Captain. He spent the next three years overseas in the Tactical Air Command in Africa, Italy and France, rising to the rank of Lt. Colonel at the time of his discharge in May 1945. Harold stayed in Meriden to run the company with the title of Manager. Parker's wife, Elizabeth, came in as Treasurer, pay mistress and protectress of the Allen family's financial interests. Harold had recently hired John Peckham from Princeton as his assistant. But John was drafted in 1942 and was gone for the duration, as were a number of shop personnel.

Labor soon became scarce. Paper was not rationed but was controlled and allocated by the government for use in "essential" war-related purposes. These scarcities threatened the continued viability of the business. Harold scrambled to get contracts that qualified as war related so that the paper he needed could be obtained, and so he could keep on what was left of his skilled work force.

The greater part of the work done in the next four years was for defense contractors or branches of the armed services. General Electric was manufacturing a wide variety of military products, all of which needed manuals for training and maintenance. GE thus became an important customer. Perhaps the most interesting printing that Meriden did was a top-secret project for the Office of Strategic Services.[7] This was a propaganda piece masquerading as a popular Japanese magazine that purported to instruct the civilian population about personal safety when under American attack and told of "the millions of tons of bombs that would be dropped." The magazine was to be dropped over cities in the hopes that it would undermine Japanese resolve. To be authentic, it had to be poorly printed on the cheap paper that was in use in wartime Japan. For this Meriden had to work hard to lower its standards.[8]

Because of the shortage of labor, the more productive offset press was used extensively in preference to the collotype presses. Although Meriden's offset work was much inferior to its collotype for scholarly publications, the quality was adequate for the military and industrial printing that sustained

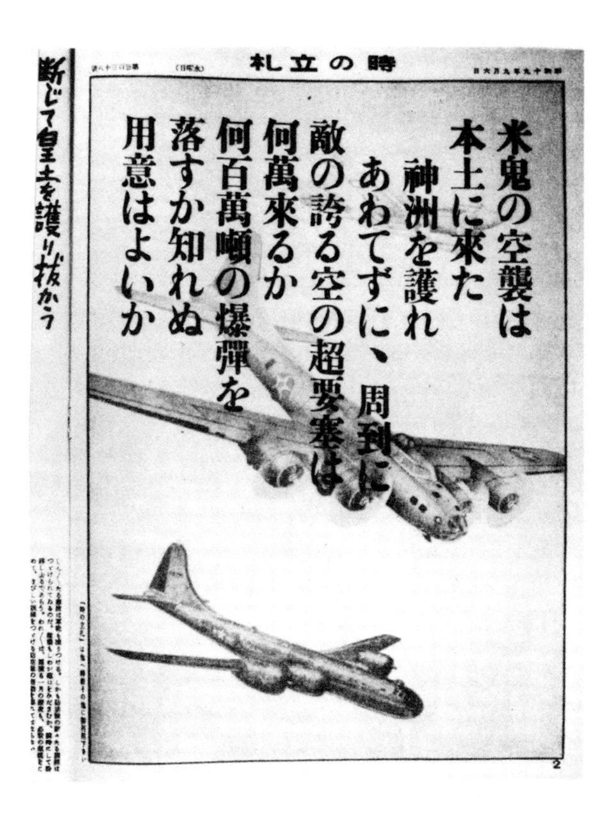

FIG. 3.2. Fake Japanese magazine printed by Meriden for the Office of Strategic Services, as reproduced in *Printing History*, No. 20, 1988, from the copy in the Meriden Gravure Archive, Beinecke Library, Yale University.

it during the war years. Indeed, Meriden was given work that another more experienced but less quality conscious offset house could not do. This had the unintended effect of strengthening Meriden's capabilities in offset and pointed the direction that it was to take in the postwar period.

When the war ended, several important decisions were made that would have a profound influence on the future of the company. Parker returned to Meriden but left the day-to-day management of the Gravure in Harold's hands while he attended to the needs of his other family-owned company, the Charles Parker Co. Secondly, it was decided that the Gravure would not go back to commercial work and would not be tethered to the fortunes of the silver manufacturing industry. The concentration was to be on scientific and fine arts work and would be diversified over many customers rather than concentrated on a few. Finally, it was decided to try to develop the company's offset capabilities to match the quality standards of its collotype work, rather than try further efforts to improve the collotype process. The Lithographic Technical Foundation played an important role in research and development of photo-offset, as did Eastman Kodak and DuPont, the principal suppliers of photosensitive materials to the printing industry. Rochester Institute of Technology, with financial and technical encouragement from Kodak, became a center of research and development activity as well as a training ground for generations of students entering graphic arts careers. Some of the advances made were appropriate only to large-volume printing, while others proved adaptable and helpful to the high-quality short-run market that Meriden pursued. Particularly helpful to Meriden was the development of pre-sensitized, smooth-surface offset plates. Previously, the printer would have to grain his own plates, then sensitize them using a whirler, which would spread a uniform thin coating of the photosensitive solution over the plate by centrifugal force. This was a tricky business. Now plates came from the manufacturer uniformly sensitized, securely wrapped to prevent premature exposure to light, and ready to expose right out of the box.

In both letterpress and offset, halftone illustrations were produced with screens that broke the image up into dots. If the screens were fine enough, the dots would not be apparent to the naked eye and the image would appear to be a smooth modulation from one tone to another. The coarseness of the screen was measured by the number of dots per linear inch, and the variability of the tone at any given point was determined by the size of the dot. Newspaper work was 85 dots per inch, which is visible to the naked eye. High-quality commercial work was 133 lines, just below the threshold of visibility. Meriden pioneered in printing in 300-line screen. Collotype was a screenless continuous-tone process although, under magnification, it had its characteristic reticulation. (See Fig 1.2) Offset could not avoid the screen, but at 300 lines per inch it gave the appearance of being screenless.

The master magenta contact screens required to do this were developed by Kodak during the war and were classified and not available to commercial printers even for war use, although Meriden seems to have gotten a screen for experimental use. As soon as the war was over and these restrictions were lifted, Meriden bought the screens and devoted itself to mastering the process. The difficulty in using them was that because of the fineness of the screen, the tolerances both in making the negatives and in printing from them were much tighter than on coarser screen work. Harold felt that Meriden had to master the 300-line screen if offset was to succeed, and he persistently pushed the shop toward that goal.

The two men in the shop who were most responsible for helping Harold to realize his ambitions were Frank Poll, the foreman of the pressroom, and Roger Bartlett, the head of the camera department. Each department had its

Fig. 3.3. Bill Liebe, pressman; Frank Poll, pressroom foreman; Jay Barefoot, press helper. (Photo, Robert Hennessey)

pride and, if something did not go well, it was assumed to be the fault of the other fellows. But Bartlett and Poll were resourceful and talented men and worked cooperatively with one another to reach the common goal. This was to do offset printing that had all of the advantages of collotype, without the disadvantages. It didn't go easily, but after many false starts they developed the practices that achieved the objective. Once the personnel had mastered fine screen work, it became routine and easier to do than coarse screen work.

Harold recalled that originally "it did not occur to us that a printer should attempt to control the quality of the copy; we were in the business of making reproductions. But it started us thinking, and when we later got photographs for the copy we tried to get an opinion from the author as to their adequacy;

FIG. 3.4. Roger Bartlett, supervising cameraman, 1962. (Photo, Carl F. Zahn. Hugo Family Papers)

then we tried to go one step further to get the negatives from which these photographs were made so that we could make our own positives and possibly get more detail and more tonal scale out of the negatives."[9]

Offset was improved by running the presses slower than was customary, which was still much faster than the collotype presses were run. The slower press speeds permitted the use of stiffer, tackier inks than were commonly used in offset. This resulted in a more faithful image reproduction. Careful attention was given to making sure that the ink did not emulsify and result in a weak, grayed-out appearance that was a common and justified criticism of offset printing at the time. It was Meriden's innovative practice to pack the blanket cylinder to increase the pressure and the ink coverage in the dark areas of the illustration while holding open fine detail in the lighter areas. This was a common practice in letterpress printing but was unheard of in offset printing. The presses at Meriden printed one side of the sheet at a time, allowed it to dry overnight and then backed up the sheet the next day. Perfecting presses were available that could print both sides of the sheet at one pass, but the ability to run the ink full and rich was compromised in doing so. For the relatively short-run work, usually less than ten thousand copies, that Meriden was doing, the extra press setup and makeready time required to perfect the sheet negated the time saved by printing both sides of the sheet at once.

Additional single-color Harris offset presses were added in the late forties and fifties to make a total of four in addition to the three collotype presses that had been kept on. The offset presses were faster, less labor-intensive, more consistent throughout the press run and more capable of printing color in close register. Collotype could print only on uncoated paper; offset could print on either coated or uncoated as best suited the job. Offset had a simpler plate-making system, more durable aluminum plates and a capacity for longer runs without having to make new plates. In the postwar period, the expansion of higher education increased the demand for scholarly publications. Even books on very narrow and esoteric subjects were now produced in thousands of copies where previously hundreds of copies would have been more than enough.

When the particular tonality and richness of collotype had to be emulated in offset, it was necessary to print the images in two colors as duotones. Two negatives were made: a rather contrasted negative was made for the black printer, and a separate negative at a different screen angle with a flatter tonal range was made for the second impression. This would be printed in a cool gray or a warm brown or something in between. There was no standard, "one color fits all" ink. For each job the ink was custom mixed as appropriate and its ingredients carefully recorded for reference in the event of a reprint.

Running multicolor work on a one-color press meant running "dry trap," that is to say, the ink on the first pass through the press was allowed to dry, usually overnight, before the second impression was put down. The paper unavoidably absorbed some moisture from the press dampening system, which caused the paper to expand. To minimize this expansion and to hold register, the paper grain had to be with the long dimension of the sheet. At the same time the bind dimension of the page had to be along the grain so that the pages would fold smoothly in the bindery and open freely in the hands of the reader. The solution to this problem was to run a typical 8-½ x 11 inch trim size, six pages out of a 23 x 35 inch sheet instead of eight pages out. This satisfied the demands of both the press and the bindery, but it added to the cost of printing and binding, and usually added to the cost of paper as well.

Meriden did not make its own four-color process separations. It would subcontract this work to a firm with specialized expertise in dot etching for color correction. Its favorite source was Rainbows Inc., a small family-owned business up in Hazardville, Connecticut. Rainbows was run by Will Mangini, a consummate craftsman and exuberant personality who was willing to give the work the extra care and personal attention that Meriden required.

Printing multicolor work one color at a time was an adventure because you couldn't see at the outset what the final result would be. Experience was helpful but not a total solution. The better practice was to run press proofs that were a dress rehearsal for the actual press run. This helped to anticipate any problems that might occur before things got too far along, and it was reassuring to the customer. Many printers did this, but most did it on a proofing press and not a production press. Meriden felt that the only really valid test was one done on a production press, and it was usually the actual press and press crew that were to run the job. It was Meriden's practice to press proof all four-color process work and a representative sampling of each duotone job.

Poll served as pressroom foreman for some thirty years. He was intense, resourceful, had a sharp eye for print quality and an instinct for leadership. He had a great respect for Harold's commitment to quality, and Harold fully appreciated his talent and his contribution to the success of the company. Poll could tell just by listening that the presses were running smoothly. He was like the mother of several infant children who could distinguish between the cries of each and sense the meaning of the cry. This was before OSHA began insisting that pressroom personnel wear earplugs to protect against hearing loss. Poll hovered over his press crews and was quick to intercede if he felt it necessary. He took a managerial attitude toward his work and was always looking ahead, anxious to keep the presses productive every minute of each working day. He was constantly thinking of how he could meet production schedules with the absolute minimum of changeover time between jobs.

When Poll retired, his successor, recruited from elsewhere in the industry, though competent, was not as managerial or energetic. As a result, more of the responsibility for press quality fell on the pressmen themselves, and the production manager assumed the scheduling worries.

Bartlett, in the camera room, was more relaxed and less high-strung than Poll, and not as managerial. In addition to supervising the work of the other cameramen he did camera work himself, including many of the most challenging jobs. When there was camera work to be done at the off-premises camera installations in Boston or New York, it was usually his assignment.

Paper and its Problems

The paper Meriden used had to be of especially high quality. Uncoated stock had to have a high pick resistance, that is, the paper fibers had to be strongly bonded to one another so that the stiff, tacky ink, applied under great pressure, would not pull the fibers loose. For coated papers, the coating had to be strongly bonded to the base stock. The presswork practices at Meriden made greater demands on the paper than did other printers. Most manufacturers were content to make paper that met common industry standards, but which did not always meet Meriden's more stringent requirements. One solution to this problem was the development of Meriden's proprietary house sheet, SN Text. This paper took its name from the merchant that initially assisted in its development, the Stevens-Nelson Paper Co., which later became Andrews-Nelson-Whitehead. The paper was for a long time produced at the Curtis Paper Co. of Newark, Delaware. SN Text was manufactured in carload quantities in 70 and 80 pound text weights. It was a slightly warm, creamy white sheet with a hard finish that held the ink up on the surface so as to enhance the richness of the illustrations. The Meriden account at Andrews-Nelson was handled by the veteran paper salesman Herbert Farrier. Over the course of his long association with Meriden, he became one of Harold's most devoted friends. If a shipment of paper was in some way defective, this problem only became apparent when the job got on press. This did not happen frequently, but when it did it was always, by application of Murphy's Law, on a job that was in a desperate rush to get printed, bound and delivered. In one such situation, Farrier remarked that papermaking practice had changed. It had become more a matter of scientific measurement and numerical control and less influenced by discernment and instinct of skilled, experienced workers. Harold thought that the achievement of quality was an art, not a science, and he regretted this trend.

A related matter that occasionally came up was the use of ink densitometers in the pressroom. These were optical devices that measured the amount of ink on the color control bars that ran on the bottom edge of the sheet. They were useful in maintaining consistency throughout the run. But in Harold's view the final arbiter of quality would always be the sensitive human eye. His sympathies were firmly on the side of craftsmanship rather than gadgetry.

Meriden also used other nonproprietary Curtis papers, particularly Curtis Rag, and various cover stocks made in a variety of colors and finishes. Other sheets that Meriden ran frequently and successfully were Mohawk Superfine Text made in Cohoes, New York; the Monadnock line made in Bennington, New Hampshire; and the Strathmore cover papers made in Woronoco, Massachusetts.

When coated stocks were called for, Meriden's preference was for the coated papers made by the S. D. Warren Co., particularly Warren's Lustro Offset that was available in both gloss and dull finishes. Occasionally an especially deluxe job was done on handmade paper but, if so, Meriden insisted that the sheet be impregnated with sizing that would securely bind the fibers, and that would resist moisture that transferred to the paper from the press dampening system. In the printing for the illustrations for the Gehenna Press *Francesco Laurana* this had an unintended beneficial effect. The paper was Tovil, an English handmade sheet. The sizing was applied only to the image area and not to the entire sheet. It gave an appropriately warmer cast to the reproductions, which were much admired. Several subsequent publications also made use of this discovery.

The paper manufacturers just mentioned produced their paper on Foudrinier machines the length of a football field. But for all that, they were still rather small-scale boutique manufacturers compared to the large-volume paper companies supplying mass-market publishers and advertisers. Working with Meriden was, for the paper manufacturers, sometimes a hassle. It was demanding, but it was also a source of pride and prestige for the manufacturer that made the effort worthwhile.

In the 1950s research libraries became aware of the deterioration of their nineteenth-century holdings due to the acidity of the paper. Libraries campaigned for the adoption of acid-free papers henceforth. The American Association of University Presses took up the cause, as did some trade publishers. Paperback publishers, particularly those that issued popular fiction, resisted the movement on the grounds that it added to costs, and that their books were ephemeral anyway. Meriden, mindful that its books would have long-term value and relevancy, was an early and active supporter of the movement.

"Can We Work from the Originals?"

It was Harold's conviction that the best and most accurate reproductions were those made directly from the original to a screened halftone negative. This eliminated making a continuous-tone photographic negative and then making a continuous-tone photographic print from which the ultimate halftone negative was made. Too much of the character and subtlety of the original was lost in these intermediate steps. Shooting from the original gave the reproduction freshness and authenticity. The process cameras used in the printing industry had lenses that had no depth of field. They could only handle flat, two-dimensional copy. This meant that three-dimensional objects such as sculpture could not be photographed directly. But images on paper—drawings, prints, art photographs and printed pages or documents—certainly could be. If an art historian was publishing a paper trying to establish that a drawing was not by Rembrandt but rather by one of his followers, the illustration had to be detailed and accurate enough to make his contention convincing. Muddy third-generation glossy photos might do for newspaper use, but these photos were seldom of sufficient quality to support a serious and possibly disputed scholarly argument.

The Museum of Fine Arts, Boston became convinced of the merits of Harold's practice on a publication of William Blake's watercolor drawings. Carl Zahn described the situation.

> "This was in 1957, a year after I joined the staff, and there was plenty of convincing to be done in order to obtain permission for the watercolors to travel from Boston to Meriden. Several subjects were reproduced from regular glossy silverprints and also from the originals. No further comment needed! Off they all went to Meriden and, in testimony to Harold's loving care and handling, in twenty-two years of my experience every work of art entrusted to him has been returned unblemished to the Museum." [10]

In other situations the advantage of shooting from the original was that by doing so, the image could be reproduced as a line illustration with greater contrast and clarity than would be possible if reproduced as a halftone from a glossy photograph.

One important example of the benefits of shooting from the original was the *Massachusetts House Journals*. These were the printed record of the actions of the Massachusetts legislature from 1715 to 1776. No institution had a complete run of the *Journals*. The Massachusetts Historical Society had most, and what they lacked was, fortunately, available at Harvard or the American Antiquarian Society.

The Commonwealth of Massachusetts gave the MHS subsidies to supplement MHS endowed funds needed to publish one volume of the *Journal* each year. Eventually the complete series became available to learned institutions. At the start, the text of each volume was painstakingly reset in type, complete with the typographic conventions common to eighteenth-century printing. Harold persuaded the MHS that it would be better and less expensive to photograph the pages of the original book and reproduce them as line illustrations. All the

VOTES

Of the House of Representatives.

Lunæ 30. Die Maii, A. D. 1748.

Petition of Benj. Rider.

A Petition of *Benjamin Rider* in Behalf of the second Precinct in *Plymouth*, shewing that for three Years past Collectors have been chosen to collect the ministerial Rates in said Parish, but have never qualified themselves by taking the Oaths, neither have they gathered the Rates; they therefore pray this Court would impower said Collectors to qualify themselves, the Time being elapsed notwithstanding, or that they may be otherways relieved.

Read and committed to Col. *Otis*, Capt. *Little*, and Col. *Choate*, to draw a proper Resolve in Answer thereto.

Representatives returned from several Towns.

Israel Williams, Esq; returned Representative from *Hatfield*,

Joseph Freeman, Esq; returned Representative from *Harwich*,

Humphry Chadburne, Esq; returned Representative from *Berwick*, making their Appearance in the House;

Ordered, That Capt. *Bragdon* attend on the said Gentlemen to *Jacob Wendell*, Esq; and others, appointed by his Excellency and impowred by Dedimus to administer the Oaths of Allegiance &c. to the respective Members.

Which Capt. *Bragdon* returned he saw done.

And then the said Gentlemen took their Seats in the House.

Memorial of Eben. Prout.

A Memorial of *Ebenezer Prout* a Commissary in the late Expedition against *Cape-Breton*, shewing that in *April* last the Court ordered the Commissary-General to settle his Accounts, agreable to an Order of the 2d of *August* 1745, by allowing and paying him for the Rum, Molasses and Beer, which he should make appear to have actually delivered over and above what he has received; but that the Commissary-General don't see fit to allow him but *ten Shillings* per Gallon for Molasses, and *fourteen Shillings* per Gallon for Rum; by which he shall be a great Sufferer: The Memorialist also suggests there was an Order of the Court *February* 15th 1745, for paying the Men their short Allowance [16]

FIG. 3.5A. *Journal of the Massachusetts House of Representatives* for the session of May 1748, typeset and printed by Wright & Potter Printing Co., Boston. Published as Volume 25 of the series by the Massachusetts Historical Society, 1950. (Illustration courtesy of the Massachusetts Historical Society)

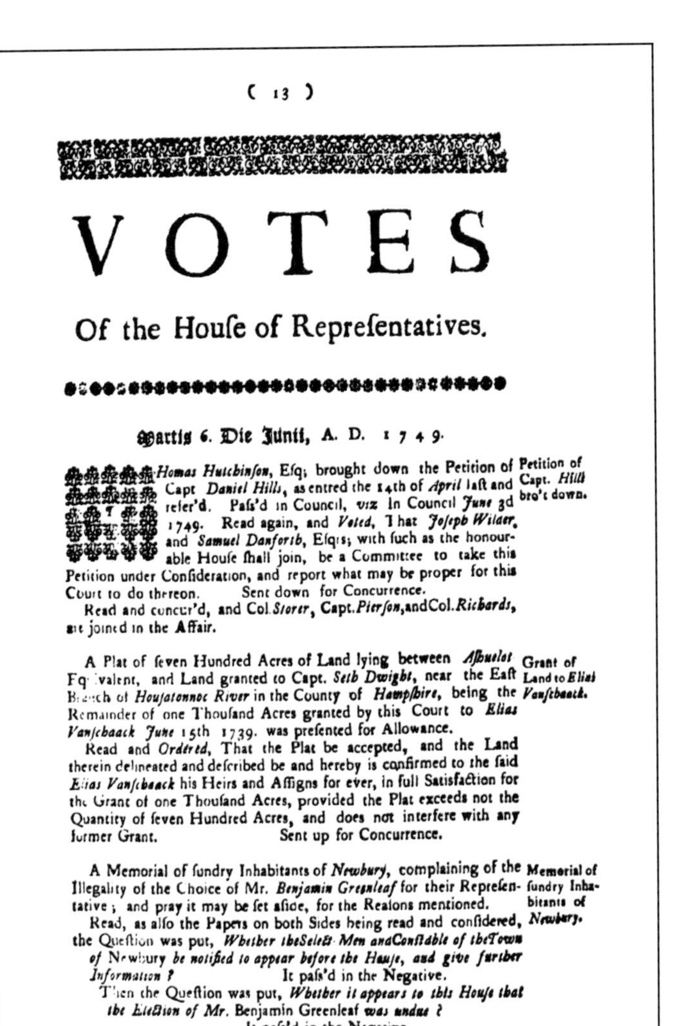

(13)

VOTES

Of the Houſe of Repreſentatives.

Martis 6. Die Junii, A. D. 1749.

Petition of Capt. Hills bro't down.

THomas *Hutchinſon*, Eſq; brought down the Petition of Capt *Daniel Hills*, as entred the 14th of *April* laſt and refer'd. Paſs'd in Council, *viz* In Council *June* 3d 1749. Read again, and *Voted*, That *Joſeph Wilder*, and *Samuel Danforth*, Eſqrs; with ſuch as the honourable Houſe ſhall join, be a Committee to take this Petition under Conſideration, and report what may be proper for this Court to do thereon. Sent down for Concurrence.

Read and concur'd, and Col. *Storer*, Capt. *Pierſon*, and Col. *Richards*, are joined in the Affair.

Grant of Land to Elias Vanſcbaack.

A Plat of ſeven Hundred Acres of Land lying between *Aſhuelot* Equivalent, and Land granted to Capt. *Seth Dwight*, near the Eaſt Branch of *Houſatonnoc River* in the County of *Hampſhire*, being the Remainder of one Thouſand Acres granted by this Court to *Elias Vanſcbaack June* 15th 1739. was preſented for Allowance.

Read and *Ordered*, That the Plat be accepted, and the Land therein delineated and deſcribed be and hereby is confirmed to the ſaid *Elias Vanſcbaack* his Heirs and Aſſigns for ever, in full Satisfaction for the Grant of one Thouſand Acres, provided the Plat exceeds not the Quantity of ſeven Hundred Acres, and does not interfere with any former Grant. Sent up for Concurrence.

Memorial of ſundry Inhabitants of Newbury.

A Memorial of ſundry Inhabitants of *Newbury*, complaining of the Illegality of the Choice of Mr. *Benjamin Greenleaf* for their Repreſentative; and pray it may be ſet aſide, for the Reaſons mentioned.

Read, as alſo the Papers on both Sides being read and conſidered, the Queſtion was put, *Whether the Select Men and Conſtable of the Town of* Newbury *be notified to appear before the Houſe, and give further Information?* It paſs'd in the Negative.

Then the Queſtion was put, *Whether it appears to this Houſe that the Election of Mr.* Benjamin Greenleaf *was undue?*

It paſs'd in the Negative.

Then the Houſe adjourn'd 'till to Morrow Morning ten o'Clock.

D Mercurii

FIG. 3.5B. *Journal of the Massachusetts House of Representatives* for the session of June 1749, photographed from the original and printed in facsimile by Meriden Gravure. Published as Volume 26 of the series by the Massachusetts Historical Society, 1951. (Illustration courtesy of the Massachusetts Historical Society)

laborious typesetting, proofreading and makeup would be eliminated without any loss of content. Harold's proposal was accepted, almost certainly after cost estimates and sample press proofs demonstrated the clear advantages of his suggestion. Stephen Riley, the Director of the MHS, became a convert to and subsequently an apostle of the Hugo philosophy. He has been quoted as saying, "Harold, the trouble with you is that you are always making us librarians do what we should have done long ago anyway." [11]

Another beneficiary of this practice was the *Harvard College Library Catalogue of French Sixteenth Century Books*, published by the Harvard University Press for the Library in 1964. The catalogue of 1060 titles, all of them illustrated, came to 728 pages in two volumes. All the books came to Meriden, fifty at a time. This permitted most of the illustrations to be rendered as fine-line images.

Meriden had printed the illustrations for Philip Hofer's *Baroque Book Illustration*, published by Harvard University Press in 1951. There were 150 illustrations, and the majority of them were reproduced in fine line with great clarity and no loss of detail. In retrospect it seems possible that Hofer may have had in mind using this publication as a trial. If Meriden's reproduction technique was successful, it could be applied to the much more ambitious catalogue of the French books. Vivian Ridler, Printer to Oxford University, was among those greatly impressed with Meriden's offset work on *Baroque Book Illustration.* He conceded that it far surpassed what was then being done on similar publications in the U. K.[12]

In his review of the *Catalogue of French Sixteenth Century Books*, William A. Bostick wrote,[13]

> "technical virtuosity is shown in the line reproductions (i.e. with no halftone screen). The negatives and the presswork bring out every nuance of the original type and illustrations. Where the sixteenth-century pressman was careful in inking and printing, none of the sharpness is lost in the 'Kiss' impression of Meriden's photo-lithography." He goes on to say, "On a par with the fidelity of type reproduction in the Harvard catalogues is the duplication of woodcuts and engravings. Most of these are reproduced along with the type used on the original page, thus showing the graphic harmony achieved by the French Printers. Most of them even when reduced to half size or smaller still retain the clarity of the original printing. This is easier for woodcuts than for engravings, and the results achieved in the latter are truly amazing. In addition handwritten notes and even the offset of ink from one original wet page to another (indicating a proof copy rushed from the press) are reproduced without the crutch of a half-

FIG. 3.6. Jacket for *Baroque Book Illustration*, Harvard University Press, 1951. Calligraphy by Rudolph Ruzicka.

> tone screen." "For the Harvard catalogue the publishers most wisely took their reproduction problem to that *maître du fac-similé*, that practicer of lithographic legerdemain—Harold Hugo of the Meriden Gravure Company." [13]

This catalogue was the accomplishment of Ruth Mortimer, whose knowledge of sixteenth-century printing and publishing practices and meticulous bibliographical descriptions made this a classic of its kind. It was followed by its sister publication, Mortimer's *Catalogue of Italian Sixteenth-Century Books*, done to the same high editorial and production standards and published by Harvard in 1974.

ΑΠΠΙΑΝΟΥ ΑΛΕΞΑΝΔΡΕΩΣ
ΡΩΜΑΙΚΩΝ

Κελτική	Συριακή
Λιβυκὴ ἢ Καρχηδονική	Παρθική
Ἰλλυρική	Μιθριδάτειος Ἐμφυλίων ε΄.

APPIANI ALEXANDRINI ROMANARVM HISTORIARVM

Celtica	Syriaca
Libyca, vel Carthaginensis	Parthica
Illyrica	Mithridatica Ciuilis, quinque libris distincta.

EX BIBLIOTHECA REGIA.

Βασιλεῖ τ' ἀγαθῷ κρατερῷ τ' αἰχμητῇ.

LVTETIAE
Typis Regiis,
Cura ac diligentia Caroli Stephani.
M. D. L. I.

Cum priuilegio Regis.

First copy: 12½ × 8¼″. Late sixteenth-century red morocco, gilt plain fillet borders, on the spine the gilt arms and device of cardinal Charles III de Bourbon, archbishop of Rouen and a cousin to Henri IV (Olivier-Hermal, plate 2617, nos. 1–2), g.e., with 1734 autograph of François Petit and plate of Sir Francis Boileau (library of Lord Nugent). With this is bound a Latin translation of Appianus by Sigmund Gelen, Caelius Secundus Curio, and Pier Candido Decembrio, printed at Basel, H. Frobenius and N. Episcopius, 1554. Most of the Bourbon library is now in the Bibliothèque Nationale. Second copy: 13¼ × 8¾″. Eighteenth-century mottled calf, gilt plain fillet borders, ornaments on spine, brown morocco label, sprinkled edges. Also at Harvard is a copy with the autograph and ms. annotations of Hugo Grotius and another from the library of Sir Kenelm Digby, with his arms on the binding.

Collation: A^8, B–Z^6, Aa–Ii^6, Kk^4; 198 leaves.

Ref: Brunet, I.356; Renouard, *Estienne*, p. [102], no. 4; Armstrong, *Estienne*, p. 223–224; Lau, *Estienne*, p. 67; Updike, vol. 1, p. 237–238, leaf A6^r reprod. fig. 171; not in Murray, Rothschild.

30 APPIANUS. *Des gverres des Romains livres xi* (Trans. C. de Seyssel from Latin of P. C. Decembrio). f°. 2 pts. in 1 vol. Paris, F. Prévost, for A. L'Angelier, 1580 (colophon: 1569).

APPIAN ALEXANDRIN, HISTORIEN
Grec,

DES GVERRES DES ROMAINS LIVRES XI.
TRADVICTS EN FRANÇOIS PAR FEV MAISTRE CLAVDE DE SEYSSEL, premierement Euesque de Marseille, & depuis Archeuesque de Thurin.

PLVS Y SONT ADIOVSTEZ DEVX LIVRES, TRADVICTS DE GREC EN Langue Françoise, par le Seigneur des Auenelles.

REVEV ET CORRIGE DE NOVVEAV.

A PARIS,
Pour Abel l'Angelier Libraire, tenant sa boutique au premier Pillier de la grand' Salle du Palais.
M. D. LXXX.

L'Angelier's device (Renouard 549) on the title-page. The device is in two parts; figures of angels are on a separate block into which L'Angelier's medallion is inserted. The device of Pierre Du Pré (Renouard 268, this volume cited) is on the special title-page of part 2 and is repeated at the end of that part, leaf e8^v. Unillustrated. Foliated headpiece with birds, arabesque tailpieces. Foliated and grotesque initials in three sizes, including an L and a C from Benoît Prévost's starred alphabet and an E copied from the Greek alphabet designed for Robert Estienne (see Nos. 270, 219). Smaller initials from various sets. Roman and italic letter, small italic marginalia.

APPIANUS 37

FIG. 3.7. Page from Ruth Mortimer, *French 16th Century Books in the Harvard College Library Department of Printing and Graphic Arts.* Cambridge: The Harvard University Press, 1964.

There were a considerable number of publications in which manuscripts were shot directly from the originals. These were often literary manuscripts in the author's own hand showing corrections and revisions, which gave fascinating insight into the author's creative process, probably more than the author would have wanted, had he or she known. Walt Whitman was a fertile field for manuscript study. *Leaves of Grass* went through many editions, each one enlarged and revised from the previous edition by the poet. Meriden did a facsimile of the first (1855) edition from Yale's Beinecke Library copy for the Eakins Press in 1966. This was followed by a facsimile of Walt Whitman's "Blue Book," the 1861 edition of *Leaves of Grass* with his manuscript additions and revisions, which is in the collection of the New York Public Library. It was published by the NYPL in 1968. Meriden produced *Walt Whitman's Autograph Revision of the Analysis of Leaves of Grass* for New York University Press (1974), which was Whitman's self-promoting revisions to an admiring biography of him by Dr. Richard Bucke.

Whitman's contemporary, Emily Dickinson, was given the full Meriden facsimile treatment in Ralph Franklin's edition of *The Manuscript Books of Emily Dickinson, A Facsimile Edition*, published by the Harvard University Press (1981). Meriden did a facsimile of the author's manuscript of Stephen Crane's *Red Badge of Courage*, edited by Fredson Bowers for Bruccoli-Clarke Books, published by NCR/Microcard Editions in 1973. Ten years later, Gale Research published the facsimile of the working copy manuscript of *Adventures of Huckleberry Finn* for Bruccoli-Clarke Books. The manuscript, belonging to the Buffalo and Erie County Library, was not complete but nevertheless amounted to 696 pages in Twain's clear and legible hand. Although not specifically crediting Meriden it was, as noted in the two-volume publication, "printed from negatives shot directly from the original."

Historical texts were also reproduced in facsimile. Among them were two sixteenth-century exploration texts, *A Spaniard in the Portuguese Indies: The Narrative of Martin Fernandez de Figueroa* and *Giles Fletcher, Of The Russe Commonwealth.* The first of these was reproduced from the printed 1512 Spanish edition in the Harvard College Library, to which was added a transcription into English and extensive notes by James B. McKenna. The second was reproduced from the first edition, printed in London in 1591. It needed no translation but contained extensive notes and commentary by Richard Pipes. Both were published by Harvard University Press in the 1960s.

The Plictho of Gioanventura Rosetti, a facsimile of a 1548 Venetian treatise on dying fabrics and tanning leather, with introduction, editorial notes and English translation by Sidney M. Edelstein and Hector C. Borghetty, was pub-

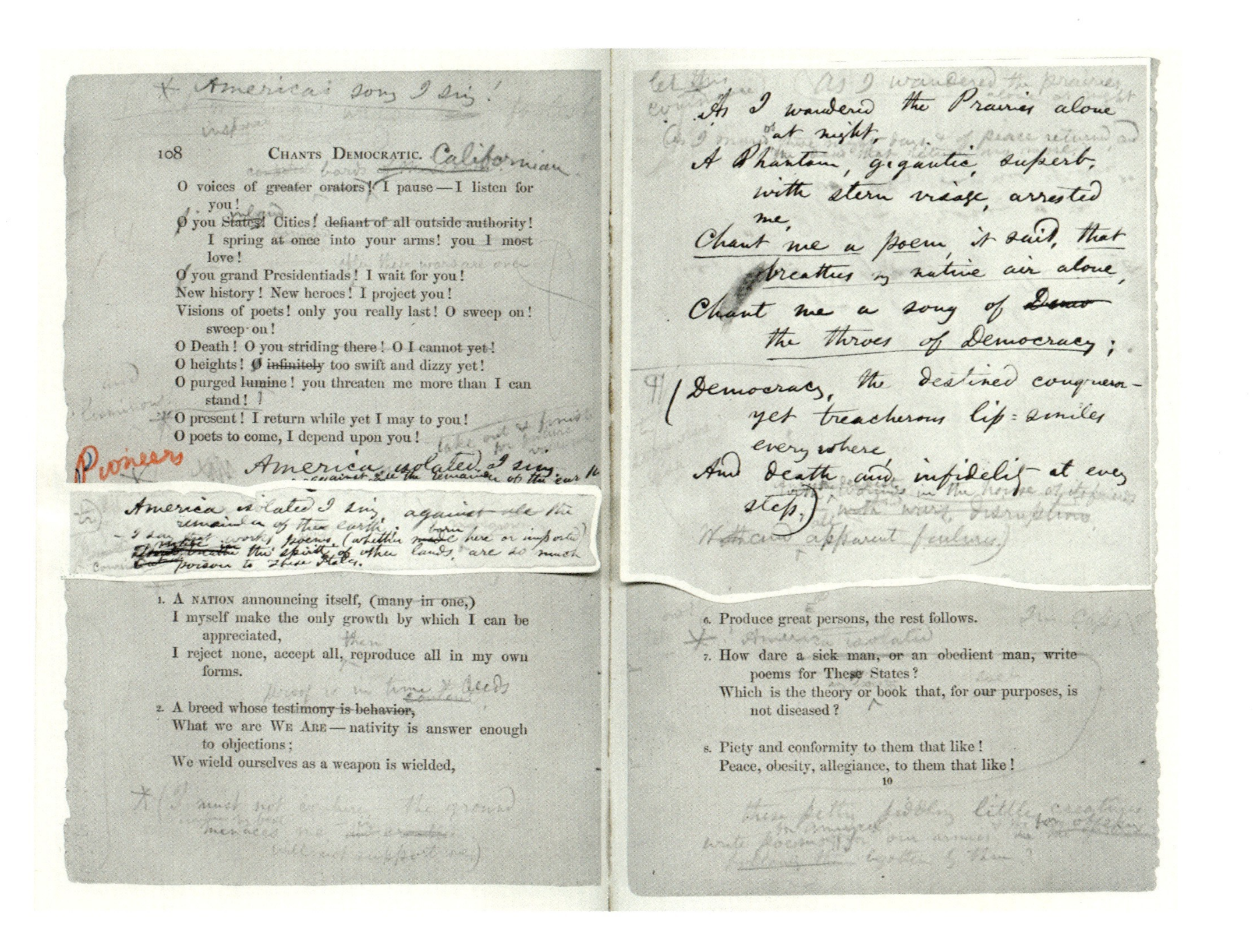
108 CHANTS DEMOCRATIC.

O voices of greater orators! I pause — I listen for you!
O you States! Cities! defiant of all outside authority! I spring at once into your arms! you I most love!
O you grand Presidentiads! I wait for you!
New history! New heroes! I project you!
Visions of poets! only you really last! O sweep on! sweep on!
O Death! O you striding there! O I cannot yet!
O heights! O infinitely too swift and dizzy yet!
O purged lumine! you threaten me more than I can stand!
O present! I return while yet I may to you!
O poets to come, I depend upon you!

1. A NATION announcing itself, (many in one,)
I myself make the only growth by which I can be appreciated,
I reject none, accept all, reproduce all in my own forms.

2. A breed whose testimony is behavior,
What we are WE ARE — nativity is answer enough to objections;
We wield ourselves as a weapon is wielded,

As I wandered the Prairies alone at night,
A Phantom, gigantic, superb, with stern visage, arrested me,
Chant me a poem, it said, that breathes my native air alone,
Chant me a song of the throes of Democracy;
(Democracy, the destined conqueror — yet treacherous lip-smiles everywhere
And death and infidelity at every step,)

6. Produce great persons, the rest follows.

7. How dare a sick man, or an obedient man, write poems for These States?
Which is the theory or book that, for our purposes, is not diseased?

8. Piety and conformity to them that like!
Peace, obesity, allegiance, to them that like!

10

FIG. 3.8. Pages from *Walt Whitman's Blue Book* with die-cut paste-overs, and a second color struck in. New York Public Library, 1968.

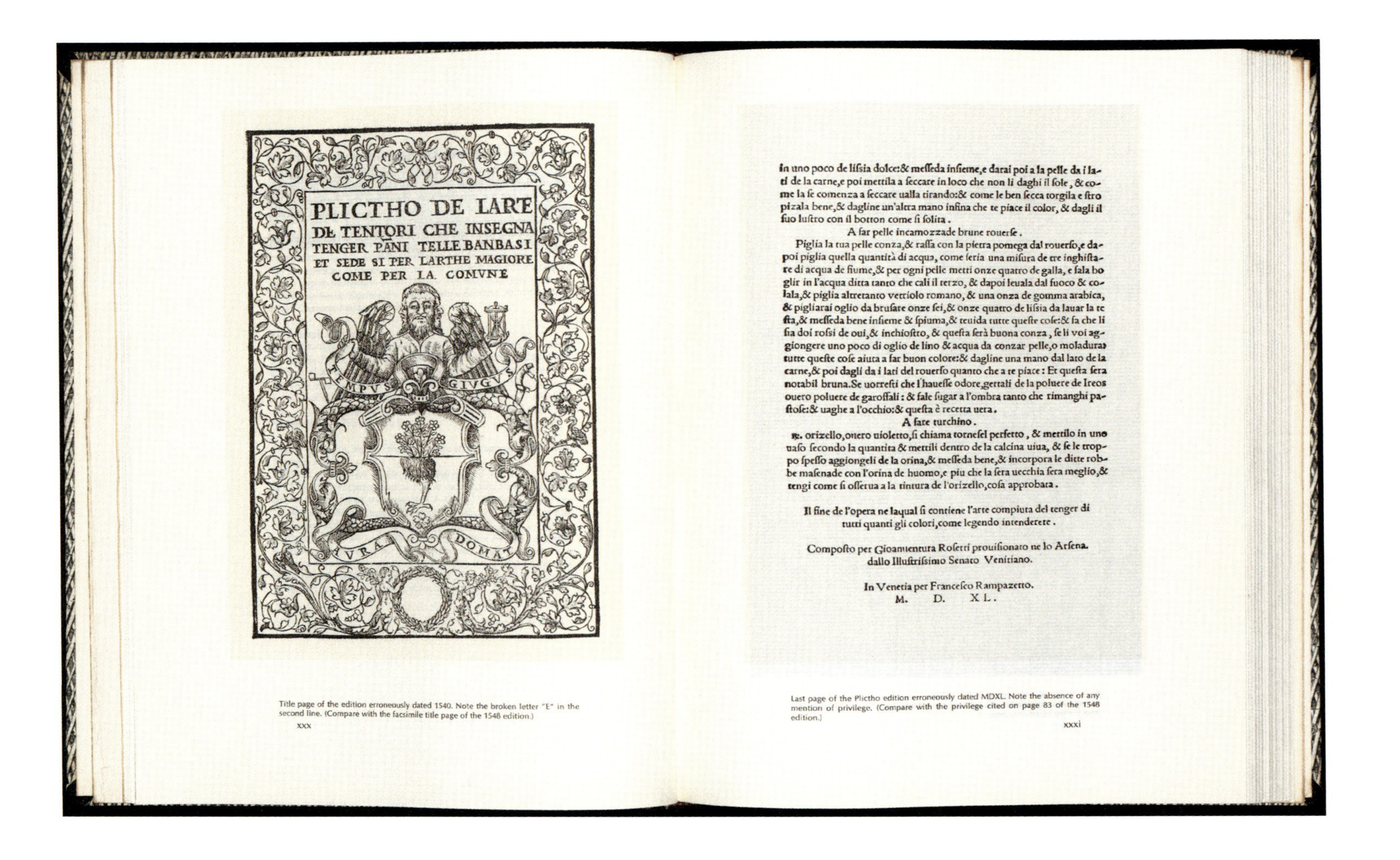

PLICTHO DE LARTE
DE TENTORI CHE INSEGNA
TENGER PANI TELLE BANBASI
ET SEDE SI PER LARTHE MAGIORE
COME PER LA COMUNE

Title page of the edition erroneously dated 1540. Note the broken letter "E" in the second line. (Compare with the facsimile title page of the 1548 edition.)

xxx

in uno poco de lisia dolce:& messeda insieme, e darai poi a la pelle da i lati de la carne, e poi mettila a seccare in loco che non li daghi il sole, & come la se comenza a seccare ualla tirando:& come le ben secca torgila e stropizala bene, & dagline un'altra mano infina che te piace il color, & dagli il suo lustro con il botton come si solita.

A far pelle incamozzade brune rouerse.

Piglia la tua pelle conza, & rassa con la pietra pomega dal rouerso, e dapoi piglia quella quantità di acqua, come seria una misura de tre inghistare di acqua de fiume, & per ogni pelle metti onze quatro de galla, e fala boglir in l'acqua ditta tanto che cali il terzo, & dapoi leuala dal fuoco & colala, & piglia altretanto vetriolo romano, & una onza de gomma arabica, & pigliarai oglio da brusare onze sei, & onze quatro de lisia da lauar la testa, & messeda bene insieme & spiuma, & teuida tutte queste cose:& fa che li sia doi rossi de oui, & inchiostro, & questa serà buona conza, se li voi aggiongere uno poco di oglio de lino & acqua da conzar pelle, o moladura: tutte queste cose aiuta a far buon colore:& dagline una mano dal lato de la carne, & poi dagli da i lati del rouerso quanto che a te piace: Et questa sera notabil bruna. Se uorresti che l'hauesse odore, gettali de la poluere de Ireos ouero poluere de garoffali: & fale sugar a l'ombra tanto che rimanghi pastose:& uaghe a l'occhio:& questa è recetta uera.

A fare turchino.

℞. orizello, ouero uioletto, si chiama tornesel perfetto, & mettilo in uno uaso secondo la quantita & mettili dentro de la calcina uiua, & se le troppo spesso aggiongeli de la orina, & messeda bene, & incorpora le ditte robbe masenade con l'orina de huomo, e piu che la sera uecchia sera meglio, & tengi come si osserua a la tintura de l'orizello, cosa approbata.

Il fine de l'opera ne laqual si contiene l'arte compiuta del tenger di tutti quanti gli colori, come legendo intenderete.

Composto per Gioanuentura Rosetti prouisionato ne lo Arsena. dallo Illustrissimo Senato Venitiano.

In Venetia per Francesco Rampazetto.
M. D. XL.

Last page of the Plictho edition erroneously dated MDXL. Note the absence of any mention of privilege. (Compare with the privilege cited on page 83 of the 1548 edition.)

xxxi

FIG. 3.9. *The Plictho of Gioanventura Rosetti* (The less sumptuous edition, on Fabriano). Introduction, editorial notes and English translation by Sidney M. Edelstein and Hector C. Borghetty. MIT Press, 1969. Approximately half size.

Fig. 3.10. Manuscript page in Beethoven's hand of the *Sonata for Violoncello and Pianoforte, Opus 69, First Movement* (somewhat reduced). Published by Columbia University Press, 1970.

lished by MIT Press in 1968. The edition of 2300 copies was printed on Fabriano 127 and hand bound by Arno Werner with paper over boards and a leather spine. An even more sumptuous deluxe edition was printed on handmade all-rag Amalfi and hand bound by Werner in genuine leather and boxed.

A number of reproductions of musical manuscripts were done at Meriden. The Pierpont Morgan Library's extensive collections were an important source of manuscript reproduction. Other sources were the New York Public Library and Yale University. These manuscripts, in widely available facsimile editions, gave musicologists insight into the practices used by the composer. They were useful to performers in their interpretation of the work. For the music-loving public to see the music in the composer's own hand gave some sense of the process of creation.

One of the most fascinating of these facsimiles was the one published by Columbia University Press of the first movement of Beethoven's well-known Opus 69 Sonata for violincello and pianoforte. Described in the introduction by Lewis Lockwood as a "preliminary final version," it is full of cross-outs, additions and scribblings as the composer kept revising his work.

Initially, if illustrations were to be shot from the originals, they had to come to Meriden. The process camera was a large permanent installation with the camera back and film holder built into a darkroom. The camera lenses in use at Meriden had a 19-inch focal length, which meant that if an image were to be reproduced at same size, it would have to be two focal lengths from the copy to the lens and two more from the lens to the film, a total of seventy-six inches. High-contrast film and a magenta contact screen were required to produce the hard-dot pattern on the film. Both of these factors added greatly to the exposure times, which were typically several minutes. Visitors to the plant whose experience with a camera was limited to a hand-held Leica with a shutter speed of a small fraction of a second were often amazed at the time and space requirements of the process camera.

Printers in general did not have a good reputation for caring for the copy that was submitted to them. Copy was frequently lost, damaged or mishandled. Harold on many occasions had to convince a skeptical curator that this would not be the case at Meriden.

Meriden had a specially built storage vault, a concrete pillbox with no interior wiring or plumbing and a steel fire door. Every original entrusted to Meriden was itemized with a valuation put on it by the owner. Meriden had insurance in effect against loss or damage, and there was a limit to the value of originals that could be held in the vault at any one time. Occasionally the valuation of material in the vault was close to bumping the insurance limit.

Fig. 3.11. Offset Camera Installation: The operator, Ed DiPersio, is adjusting the lens. The camera bellows at the left is built into the darkroom, and sheet film was loaded into the camera from the darkroom. The copyboard is to the right. (Photo, Alan Rodgers)

Acceptance of additional originals then had to be delayed until some valuables could be returned to their owners. The insurance underwriter and its local agent were always very nervous about the risks they might be incurring. This was very unusual coverage, and there was no industry-wide loss experience to guide them. In actual fact, no claims were ever made for loss or damage and the insurer pocketed a handsome stream of premiums.

Archibald Hanna, Curator of the William Robertson Coe Collection of Western Americana at Yale, wrote, "It took only a short time for the library to realize that even its most fragile and rarest possessions received the same tender care at Meriden as in the hands of Chauncey Tinker or Marjorie Wynne, This confidence was so strong that (University Librarian) Jim Babb is reputed to have told Harold, 'You can borrow anything in the library, but if you want the Gutenberg Bible you'll have to give us at least a day's notice.'"[14]

The originals that came to Meriden were never sent by mail or common carrier but were couriered by a member of the museum staff or brought to Meriden by Harold, John Peckham, James Barnett or Bill Glick. Glick remembers being told by Harold that the originals must always travel locked in the trunk of the car. Furthermore, he was warned that if he ever got in an accident, it better not be a rear-end collision.

Security had been an issue at Meriden from its earliest days. When the Gravure was doing work for rival silver companies, it was important that visiting representatives of one silver company not see the work being done for their competitors. At the time of the Second World War, there were stringent secu-

FIG. 3.12. The North wing of the plant, 1978. The protruding structure is the storage vault for valuables. The glassed-in area shown in the 1906 rendering (Fig. 1.6) has been replaced by a wood-frame roof and wall, and the production of blueprint proofs by sunlight gave way to the use of electric carbon arc lights and subsequently to mercury vapor lamps. (Photo, B. A. King)

FIG. 3.13. Illustration page from *Drawings from New York Collections II: The Seventeenth Century in Italy*, exhibition catalogue written by Felice Stampfle and Jacob Bean, published jointly by the Metropolitan Museum of Art and the Pierpont Morgan Library, 1967. The illustration shown is Ottavio Leoni, *Portrait of Settimia Manenti*, in the collection of the Morgan Library & Museum.

rity provisions to protect the original documents and the reproductions. Any waste sheets or spoilage had to be done away with lest this material fall into unauthorized hands. In the postwar years, security focused on the handling and protection of the art, books and documents entrusted to Meriden. Except during the Second World War when the government insisted on a robust security system, Meriden's security personnel were nothing more than a couple of night watchmen armed with flashlights. They were usually retirees looking for a little extra money in a not-too-strenuous job. Harold's father, Otto Hugo, held this position at one time. If there was a problem, the watchman was to call for police or fire department assistance as appropriate. Later on the watchmen were replaced by an electronic monitoring system.

Harold was not in favor of high-profile security precautions such as armed guards and chain-link fences topped with concertina wire. He favored a policy of avoiding attention to the work going on at 47 Billard St. During World War II, a Navy captain with responsibility for checking security practices of civilian contractors doing sensitive Navy work complained to Harold about the lack of chain-link fences. Harold's reported reply was that "he knew of no better way to excite the curiosity of the neighborhood and encourage employees to start reading what they were printing."[15] Harold largely had his way in this except that he agreed to arm his watchman with a revolver and instruct him never to use it.

Meriden Gravure did not solicit local job printing; it discouraged it. It did not advertise in the newspaper, did not sponsor Little League baseball teams or take ads in the high school yearbooks. Harold loved to tell the story of an English visitor who came to town and asked a policeman in downtown Meriden to direct him to the world-famous Meriden Gravure Co. The policeman had no idea where it was. He was thoroughly dressed down by the Brit, who thought that Meriden's policemen should be every bit as knowledgeable about their city as the legendary London Bobbies were about theirs.

Once the value of working from originals had been well established, Harold arranged to have process cameras installed in New York at the Metropolitan Museum of Art and in Boston at the Museum of Fine Arts. These cameras were for Meriden's exclusive use, not only for publications of those museums, but also for other nearby museums. Work for the Morgan Library was done at the Met; work for the Fogg Art Museum was done at the MFA. There were disadvantages to all parties in this arrangement. For Meriden it was the cost of putting two expensive camera installations into operation that would only be used occasionally and would for the most part lay idle. For the museums, it meant giving up valuable space in their non-public

177a. Holy Family with Saint John, the Magdalen, and Nicodemus (enlarged detail)

178a. Holy Family with Saint John, the Magdalen, and Nicodemus (enlarged detail)

FIG. 3.14. Illustration page from *Albrecht Dürer, Master Printmaker*, exhibition catalogue published in 1971 by the Museum of Fine Arts, Boston, to mark the 500th anniversary of Dürer's birth. Shown are enlarged details of two different impressions of *Holy Family with Saint John, the Magdalen and Nicodemus.*

areas that was needed for storage and other purposes. But the advantages in the quality of the reproduction and in the convenience and security to the museums made it worthwhile for them. For Meriden the arrangement greatly strengthened its connection with the museums. Eventually construction projects expanding the museums and repurposing existing space within the museums made it necessary to give up these off-premise camera facilities. It was not long afterward that digital scanners started replacing process cameras and photo-offset gave way to digital offset. But the process cameras, now become dinosaurs, had served their purpose.

It is not surprising that many of the most significant publications to come out of Meriden were reproductions of works of art on paper since these could be done from the originals. Book and manuscript reproductions have been noted, but prints and old master drawings were also a staple of Meriden's production. The Metropolitan Museum of Art did an important series of old master drawings catalogues under the curator of drawings, Jacob Bean, which were printed at Meriden. The Pierpont Morgan Library under the leadership of Felice Stampfle, was another active exhibitor and publisher of catalogues of old master drawings of their own collections or loan exhibitions. *The Master Drawings Quarterly* was edited from the Library and printed at Meriden, although it was impractical in this instance to reproduce directly from the originals.

The Museum of Fine Arts, Boston, under Eleanor Sayre and later Cliff Ackley, published important studies of the work of such major graphic artists as Dürer, Rembrandt and Goya, as well as catalogues of the work of lesser-known but important artists. The National Gallery of Art in Washington, the Art Institute of Chicago and many smaller museums would often come to Meriden for production of their prints and drawings catalogues. Among these were two small gems of the museum world, the Farnsworth Museum in Rockland, Maine, which had a fine collection of the work of the Wyeths, some of which were reproduced for sale as pictures at Meriden, and the Hill-Stead Museum in Farmington, Connecticut. Hill-Stead was the estate of the Pope family, industrialists of Hartford who collected Impressionist art and left the house and its contents to the public as a museum. It did not have changing exhibitions, but Meriden did several printings of its guidebook. In the 1980s Stephen Stinehour, then President of the Meriden Division of Meriden-Stinehour Inc., and Stephen Harvard signed up to attend a business presentation at the Sheraton Hotel and Conference Center in Farmington. They found the speakers tedious and unenlightening, and decided to skip out and go to the Hill-Stead Museum. Unbeknownst to them, the

Fig. 3.15. The National Gallery of Art's 1958 catalogue of the work of Alfred Stieglitz, designed by Bert Clarke. The text was printed letterpress at Clarke & Way, and the illustration section was printed in collotype at Meriden. Shown is Plate 15, *From An American Place, looking North*, 1931. (Photo, Beinecke Library, Yale University)

Connecticut State Police had information that thieves were planning an art heist at the Hill-Stead, and the place was staked out. Stinehour and Harvard turned up at the Museum, casually dressed without jacket and tie, driving the Stinehour Press panel truck with no logo or identification on it, and bearing Vermont plates. The trap was sprung. The police soon realized they had apprehended the wrong men and released them. If there were thieves in the area, they were deterred by the flurry of police activity. The operation concluded and the cops left with nothing to show for their effort.

The Fogg Art Museum was an active teaching museum. Harvard graduate students in the museum course were given experience in mounting exhibitions and preparing catalogues for publication, many of which were printed at Meriden. For many of these career museum professionals, their first experience with Meriden would not be their last. The art museums at Smith College, Brown University and Brandeis University, among others, were active in organizing exhibitions and issuing accompanying publications, many of which were printed at Meriden. New England's maritime history was collected, exhibited and published by the Peabody Museum of Salem, the Kendall Whaling Museum, Sharon, Massachusetts, the New Bedford Whaling Museum, Mystic Seaport and the Museum of the American China Trade, and all came to Meriden for their major publications.

At the end of World War II photography was hardly considered an art form worthy of collection and study, except by the Museum of Modern Art and a few farsighted collectors. A generation later, Ansel Adams was a household name. Museums across the country were competing with one another to build collections, and private collectors and the dealers serving them abounded. Meriden did many publications reflecting the increased attention to photography. Georgia O'Keeffe had given important collections of Stieglitz photographs to the Metropolitan Museum of Art, the Museum of Fine Arts, Boston, and the National Gallery of Art, among others. Meriden printed the catalogues of the Stieglitz photographs for each of these institutions. O'Keeffe thought well of Meriden's work and was influential in having the printing done there.

One of the first of these catalogues was printed in 1958 for the National Gallery of Art and was done in collotype. O'Keeffe wished to be present when the printing was done. Harold, mindful of the difficulties and uncertainties of the process, tried to dissuade her from coming. She was adamant; on the day that presswork was scheduled to begin, she presented herself at 47 Billard St. Harold feared that she would be so demanding that the pressmen would walk off the job in frustration. But she was on site, and Harold felt that he had no choice but to admit her to the pressroom and hope for the best. At the end of the day, he

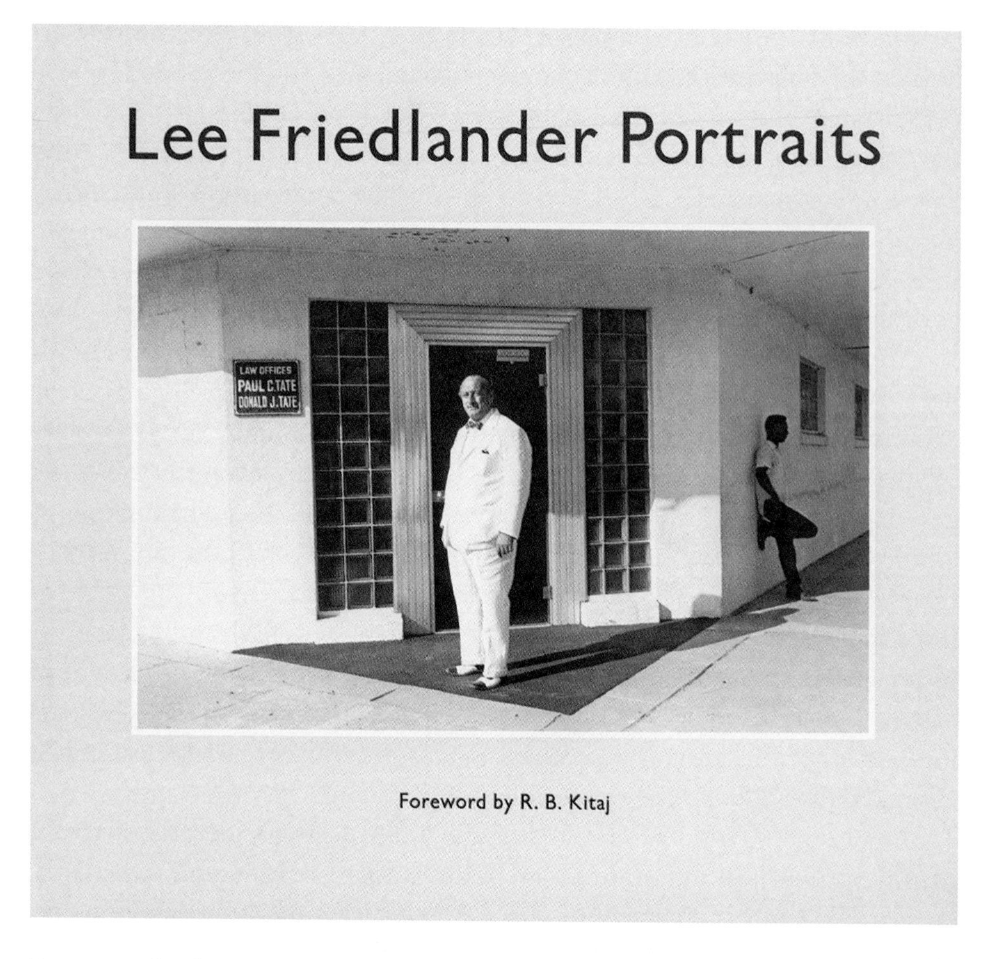

FIG. 3.16. Jacket for *Lee Friedlander Portraits*, published by New York Graphic Society, 1985. Typography and design by Howard Gralla.

took the pressman, George Zande, aside and asked how it went. Zande was one of the last and the best of the master collotype printers employed at Meriden, and he was a stutterer. His reply was "sh-sh-she has a wo-wo-wonderful eye." The job went through to completion without incident.

Harold had some further contact with her, but O'Keeffe chose to delegate most of the details of subsequent projects to her New York agent, Doris Bry, and to her favored designer, Eleanor Caponigro, working out of Santa Fe.

Other well-known photographers who got the Meriden treatment included Aaron Siskind, Harry Callahan, Walker Evans, Yousuf Karsh, Edward Steichen, Irving Penn, Paul Caponigro, Lee Friedlander and Berenice Abbott. In addition there were many publications that went through the shop devoted to particular genres of photography: architecture, environmentalism, historic preservation, social commentary, feminism, urban and rural documentaries, abstraction and formalism and the history of photography.

Meriden "Fakes"

Meriden's skill in reproducing documents with the look and feel of originals was established well back in the collotype era. These prints were always represented as reproductions, never as originals, by the publishers. They were, properly speaking, facsimiles, but in the informal parlance of the shop, they were always referred to as fakes. This usage was a proud reflection on the skill in making them, and not a reference to dishonest practice.

When the prints were matted and framed, however, and had gotten a little bit of an antique look from exposure to the sun and were eventually were passed on to a second generation of owners, they were frequently mistaken for originals. Owners were often disappointed to learn from a dealer or curator that their attractive picture was not, alas, a valuable family heirloom. Ernest Dodge, Director of the Peabody Museum of Salem, wrote, "It is a matter of record that two other of his (Harold's) Chinnery reproductions were sold and bought in a rather prominent auction as originals and from time to time his reproductions of our marine paintings are similarly mistaken."[16]

Meriden adopted the practice of printing a line in the picture where it could not be lost in the matting indicating that it was a reproduction and the year in which it was printed. This was printed in type large enough to be seen if examined at close range but not so large as to be obvious at a normal viewing distance.

Harvard, Yale, Brown, Princeton and other colleges published reproductions of their historic campus buildings for sale to alumni. Goodspeed's Book Shop offered reproductions of Currier & Ives city views of Boston. Until offset came into its own, these were printed in collotype in black and white. Frequently a blind-stamped platemark or a tint block was added to suggest that the image had been printed by an engraving process.

Hand coloring was done by the pochoir method. This involved cutting multiple acetate stencils through which the color was brushed onto the sheet. This allowed for shading the color and for soft edges. A little bit of variation and irregularity was always present to avoid a mechanical look. This hand coloring work was not done at Meriden in the postwar period. Most of it

FIG. 3.17. A Meriden Gravure "fake." A reproduction of Paul Revere's 1767 engraving of *A Westerly View of The Colledges in Cambridge New England,* from a drawing by Joseph Chadwick. It was originally a foldout plate in *Paul Revere's Engravings* by Clarence S. Brigham, published by the American Antiquarian Society in 1954, printed in collotype and hand colored. Additional sheets of this and other popular Revere prints were overrun for sale as individual prints. It is here reproduced at about one-third size. Note that in the lower left under Chadwick's name is printed "Reproduced 1954." (Courtesy, the American Antiquarian Society)

was subcontracted to the Berrien Studio, run by sisters Kathryn and Martha Berrien. Meriden's other source of hand-coloring was Maria Bittner, the widow of Harold's publisher friend, Herbert Bittner. Unlike the Berriens, who employed others to work in their atelier, Maria did all her work on her own and was limited to smaller jobs. After she lost her husband, she married Meriden's New York sales representative, Jim Barnett.

When offset became predominant, the reproductions were sometimes done in black and white, then hand-colored just as the collotype prints were. Other reproductions lent themselves to fine screen (150-line) four-color process color.

John Coolidge perpetrated a highly publicized example of Meriden's

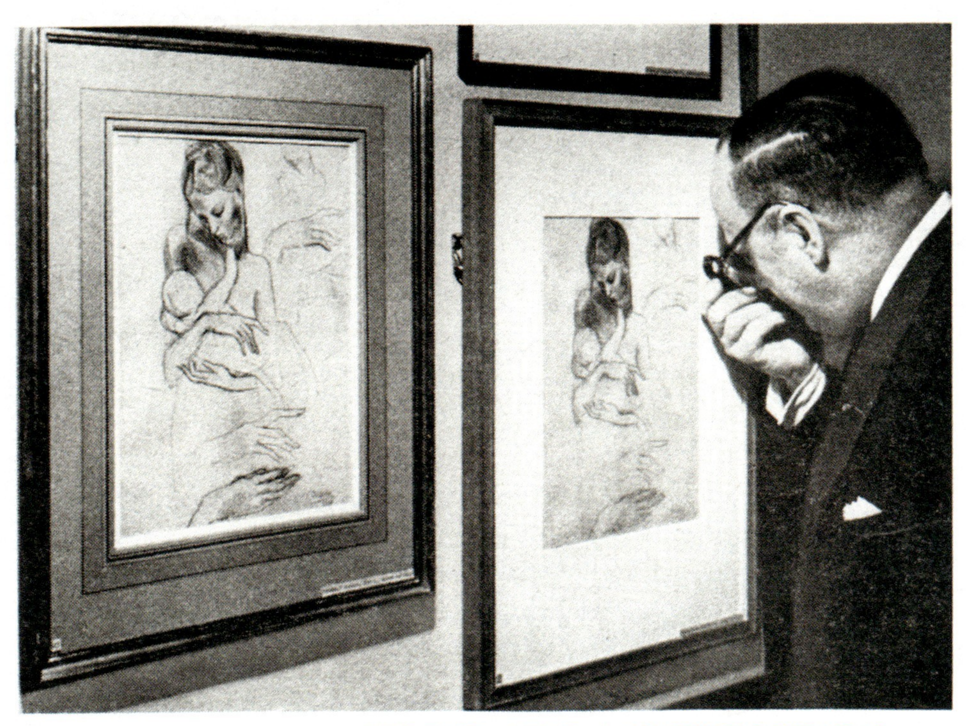

Fig. 3.18. Rorimer in *Time Magazine.*

"fakery" during his tenure as Director of the Fogg Art Museum. Meriden had done a collotype reproduction of a lovely, sensitive charcoal drawing of a Mother and Child by Picasso, a gift to Harvard from Paul J. Sachs. It was full-size, monochromatic, and was produced before Meriden had adopted the practice of identifying its facsimiles as reproductions. Coolidge assembled a small exhibition of fakes, copies and originals at the Harvard Club in New York and challenged an invited group of art experts to try and tell the difference. The Meriden fake was dressed up with a little bit of hand retouching and put into a handsome glazed and matted frame, while the original was put into a much more modest frame. Some of the experts were stumped; others weren't. In fairness to the experts it must be said that viewing a glazed and framed drawing in somewhat subdued light was not the ideal condition for determining its authenticity. The event was written up in *Time Magazine* in its typical irreverent style, and the article included a picture of James Rorimer, Director of the Metropolitan Museum of Art, studying the real and facsimile Picassos side by side.[17]

Fig. 3.19. Irving Brechlin at his collotype press, 1962. (Photo, Carl F. Zahn. Hugo Family Papers)

The End of Collotype

The collotype department was continued until 1967 despite the fact that it was a money loser. It was kept on out of loyalty to customers who wanted collotype printing and who had stood by the firm in its lean years. Another motivation was the loyalty to those employees in the department who were well along in years and could not be retrained to work in the offset department. It was also something of a loss leader that attracted potential customers. Once Meriden had a customer's interest, press proofs could be made both in offset and in collotype, and cost and time estimates submitted for both. The customer could decide. More often than not, it was in favor of offset.

FIG. 3.20. The three-story brick building seen from the rear. In the collotype era, the collotype presses were on the second floor and the letterpress department occupied the ground floor. When collotype was discontinued in 1967, the upper stories were converted to storage. The windows were then blocked in to conserve heat. (Photo, Robert Hennessey)

In 1967 one of the collotype pressmen died of a heart attack and the others, who were now past sixty-five, took their well-deserved retirement. The massive collotype presses were cut up for scrap. One was offered to the Smithsonian Institution on the condition that SI would undertake all the expenses of moving it. The offer was politely declined. The vacated space on the upper floors of the plant was used for print sample and records storage. All press production was now on the ground floor, where paper could be easily moved to and from the presses. It was at this time that the offset pressroom was expanded to allow more space for paper storage and work in process.

The end of collotype also meant the end of the letterpress department. This had been established and maintained because collotype, splendid as it was for printing illustrations on the gray scale, was poor at printing type. For some publications, particularly the scientific ones, the quality of the type was not critical, and slightly ragged collotype captions would do. For fine book work, however, a crisp, sharp impression was expected, and this

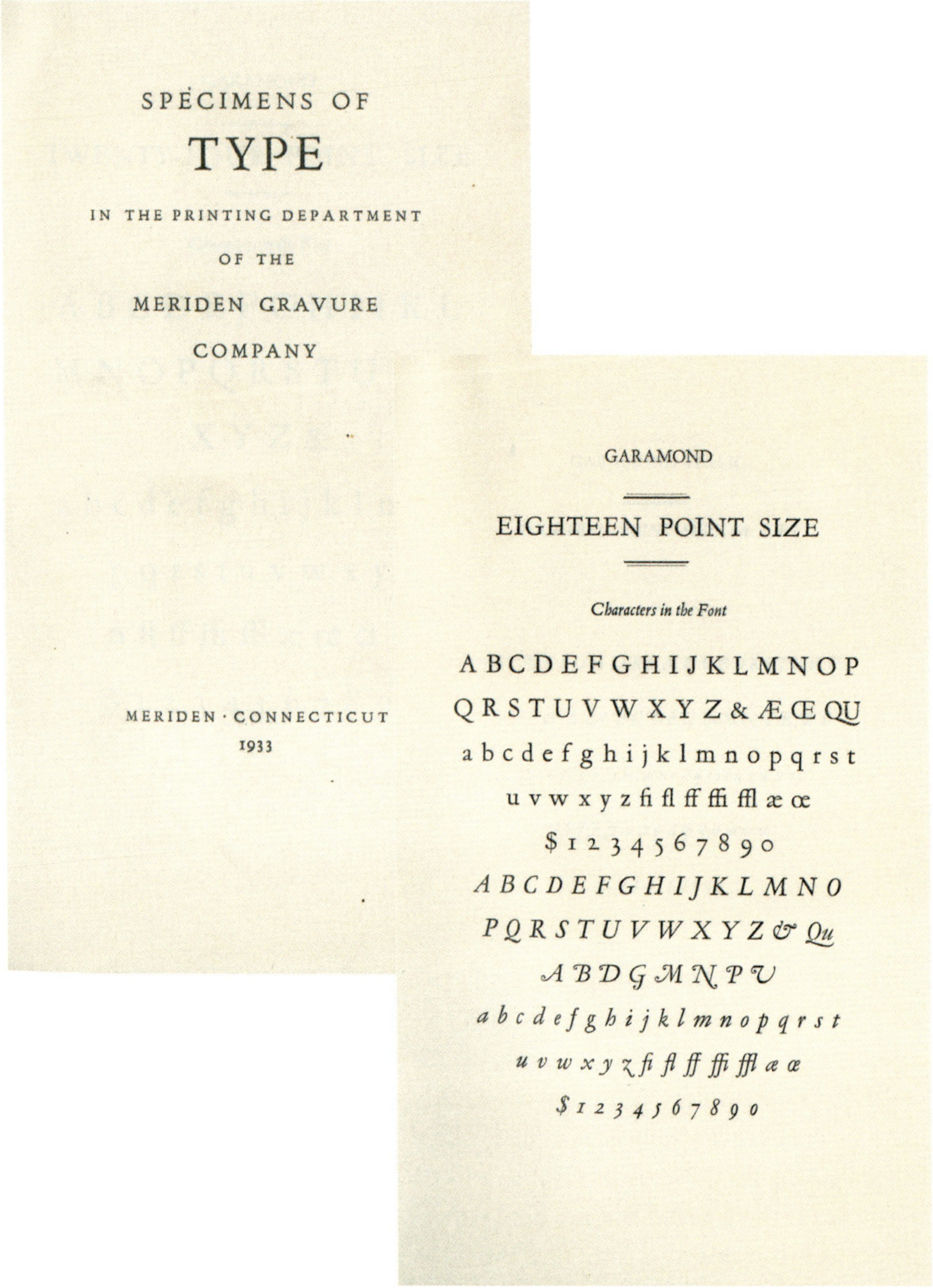

SPECIMENS OF

TYPE

IN THE PRINTING DEPARTMENT

OF THE

MERIDEN GRAVURE

COMPANY

MERIDEN · CONNECTICUT

1933

GARAMOND

EIGHTEEN POINT SIZE

Characters in the Font

A B C D E F G H I J K L M N O P

Q R S T U V W X Y Z & Æ Œ QU

a b c d e f g h i j k l m n o p q r s t

u v w x y z ﬁ ﬂ ﬀ ﬃ ﬄ æ œ

$ 1 2 3 4 5 6 7 8 9 0

A B C D E F G H I J K L M N O

P Q R S T U V W X Y Z & Qu

A B D G M N P V

a b c d e f g h i j k l m n o p q r s t

u v w x y z ﬁ ﬂ ﬀ ﬃ ﬄ æ œ

$ 1 2 3 4 5 6 7 8 9 0

Fig. 3.21. The Meriden Gravure Co. type specimen book, produced during the tenure of Gregg Anderson as head of the letterpress department.

required that the type be printed separately in letterpress on an old Babcock flatbed press that took the 25 x 38 inch collotype sheet size.

The letterpress department never had machine composition facilities but it did, in its heyday, have a good representation of type for hand composition. Garamond, Baskerville, Caslon, Fournier and Poliphilus were available in a range of sizes. Display types in roman, black letter and sans serif as well as a variety of type ornaments were in the cases. By the 1960s these fonts, worn and incomplete, were no longer useful and were cast into the melting pot.

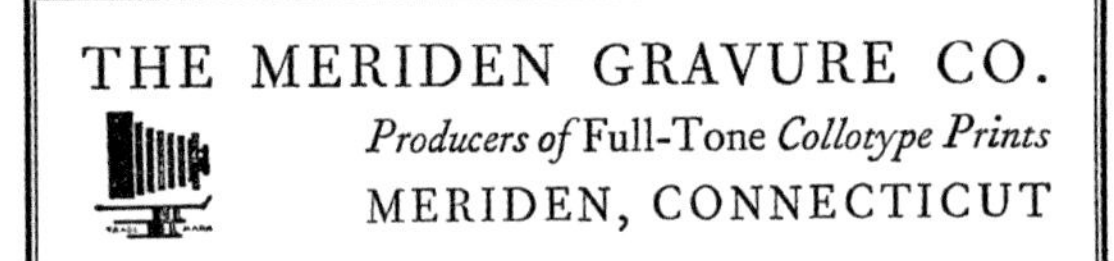

FIG. 4.1. The Home Club, built in 1903, as it appeared in 1985. With the decline of Meriden's industrial economy in the eighties, the Club could no longer sustain itself. It held the last of its legendary New Year's Day receptions on January 1, 1990, and closed its doors for good. (Photo, *Meriden Record-Journal*)

Four

Life In the Age of Offset

In the early postwar period, Meriden Gravure's largest customer was General Electric. Much of the work came from Schenectady, New York, and involved pages for massive loose-leaf parts catalogues for the generators and turbines that GE sold to the electric utilities industry. This industrial goods business was less volatile than the luxury goods business of the silver manufacturers. When dealing with an enterprise as large as GE, it was helpful to have friends at court. Over time, some of Meriden's friends had moved on in the corporate shuffle, and they were no longer able to be as helpful as they had been. This became worrisome because GE had other sources of printing including an in-house facility. Meriden needed GE more than GE needed Meriden, and that was an insecure position for Meriden.

Meriden's other important corporate customer was the Southern New England Telephone Company. SNETCO was one of the few phone companies that was independent of the Bell System prior to its breakup. For many years Meriden printed its annual report and quarterly earnings statements. These were always very "rush" and usually involved working nights and weekends to get them done. SNETCO always supplied twice as much paper for the job than was actually needed so that if disaster struck, they could reprint without delay. On several occasions this backup provision came into play. During a run, a sharp-eyed pressman noticed that the word "telephone" had been misspelled. Neither the typesetter nor the proofreaders at SNETCO had picked up the error. It was a classic case of proofreaders too familiar with the word to catch the error. The press was stopped, SNETCO officials were consulted, and it was decided to scrap the sheets, make the correction and reprint. On another occasion a reprint had to be done because the accountants decided that the financials had to be changed after they had been approved and cleared for printing.

Through the efforts of Harold and John, working from Meriden, and Jim Barnett, the resident sales representative in New York, Meriden was successful in selling to universities, research libraries, scholarly societies and art museums. The Metropolitan Museum of Art became Meriden's best customer, but its share of Meriden's annual sales was never as high as GE's had been.

Except in New York, Meriden did not employ outside salesmen in the postwar period. Monthly sales trips were made to Boston and to Princeton.

Less frequent trips were made to Philadelphia, Chicago, Washington, the Southwest and the Far West. A steady stream of customers came to Meriden to see the plant and discuss work. Customers from New York would take the 9:30 train from Grand Central, arriving in Meriden two hours later. After a tour of the shop, they would adjourn to the Home Club for a nice meal. The Home Club originated as a gentleman's club and was an important social institution in Meriden. Its membership, gained by invitation only, was made up of the leading business and professional men in the community. The Gravure entertained its customers there, and this was an important part of their experience at Meriden. The meal was generally preceded by an offer of drinks, which was seldom refused. If the beverage of choice was a martini, it was always served in a generous silver shaker manufactured by the International Silver Company. The tableware was also sterling silver of local manufacture. The Home Club was also the principal meeting place for the Columbiad Club for many years. Its facilities were attractive, and its location was a central and convenient gathering place for its members.

After the meal visitors would go back to the plant to discuss business. A late afternoon train would return them to the city, but not before they had signed Meriden's *Liber Amicitiae* (the guest book) and had helped themselves to a generous supply of samples from the conference room shelves. No business ever had less expensive or more effective advertising than these giveaways.

In addition to visiting customers, Meriden welcomed student groups. Graphic design students in Yale's graduate program and the Rhode Island School of Design undergraduate program made regular field trips to see how the printing was done. Many would return to Meriden as customers. The students were invariably fascinated with waste sheets from the pressroom. These were sheets used over and over again in the makeready process to conserve clean paper. Multiple images in a variety of colors overprinted one another in random fashion. The aesthetic of this *objet trouvé* artwork was appealing, and the sheets were eagerly snatched up and taken home as souvenirs of the visit.

Younger visitors such as school groups were discouraged. The processes were beyond their comprehension, and the machinery was a potential hazard. Further, the pressroom was not a suitable experience for young children. Every available inch of space on the walls and ceiling (often referred to as the Sistine ceiling) was covered with the centerfolds of girlie magazines. This interior decoration was not produced at the Gravure, but was readily available from a neighboring printing plant that specialized in this work. There was often some trepidation about bringing women visitors into the pressroom. But if women were offended, they were too polite to say so. By

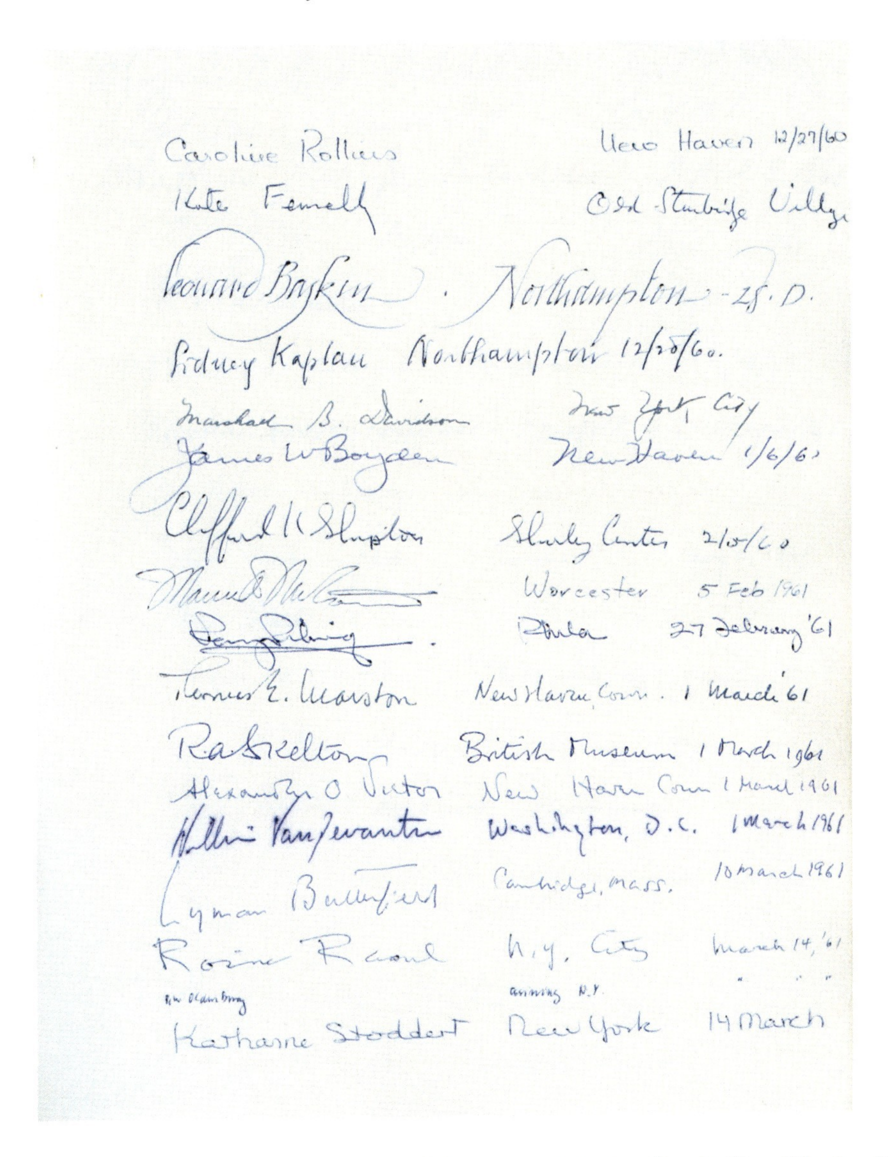

Fig. 4.2. Meriden's guest book. Signers on this page include Caroline Rollins, Yale Art Gallery; Catherine Fennelly, Editor of Publications, Old Sturbridge Village; Leonard Baskin, Smith College; Sidney Kaplan, Professor of English, University of Massachusetts, Amherst; Marshall B. Davidson, head of publications, Metropolitan Museum of Art; James W. Boyden, Manager, Yale Printing Office; Clifford K. Shipton, retired Librarian, American Antiquarian Society; Marcus McCorison, Librarian, American Antiquarian Society; Ray Riling, publisher of books on antique firearms, Philadelphia; Thomas E. Marston, Yale University Library; R. A. Skelton, Keeper of Maps, the British Museum, London; Alexander Vietor, Curator of Maps, Yale University; Willis Van Devanter, Paul Mellon Collection, National Gallery of Art, Washington, D.C.; Lyman Butterfield, Editor of the Adams Papers, Cambridge, Mass.; Rosine Raoul, publications staff, Metropolitan Museum of Art; Peter Oldenburg, Designer of Publications, Metropolitan Museum of Art; and Katherine Stoddert, Editor of Publications, Metropolitan Museum of Art. (Illustration courtesy of Stinehour Editions)

FIG. 4.3. Bob Hennessey's camera department workspace. (Photo, Robert Hennessey)

the 1970s the pin-ups had become so faded that they were taken down, and the pressroom personnel agreed not to replace them.

Meriden preferred to hire raw recruits as entry-level trainees. It was easier to train the inexperienced to the Meriden way of doing things than it was to try to retrain personnel whose habits of workmanship were not up to Meriden's. The local vocational high school, Wilcox Technical School, had a graphic arts program, and some trainees were Wilcox graduates. The Gravure funded the J. F. Allen Award, which Parker bestowed each year on the outstanding senior in the graphic arts program at Wilcox.

During the time that the collotype department remained in operation, it was difficult to recruit young people to it. The work was slow and tedious. The pressroom had to be kept at high humidity to retain the moisture in the gelatin plates. Working a collotype press in the summer was like being in a Turkish bath. Much more attractive was the air-conditioned offset pressroom with its fast, reliable precision-engineered equipment. This was where the future was. It was a common attitude in the printing industry that the

Fig. 4.4. Chien-Fei Chiang shown with his copyboard rigged up to accommodate an oversize Chinese painting. Because of his training and familiarity with the subject matter, a publication on Oriental art was usually his assignment. (Photo, Robert Hennessey)

pressroom was the place to be. But at Meriden the camera department also had its appeal. For young people fascinated with cameras and darkroom work, and who had an interest in the aesthetics of photography, the camera department was a very attractive opportunity. Seeing an image coming into being in the developing tray was a wondrous experience.

One who took advantage of this opportunity was Richard Benson, a son of John Howard Benson. The Benson family calling was stone cutting, but Richard, called Chip, was absorbed with photography. His interest included both halftone photography for reproduction and photography as an art form. His camera of choice was an 8 x 10 inch view camera, heavy and clumsy to carry about but, because of its size capable of rendering images in remarkable detail. He had a preference for warm-toned images and would make his own photo paper, using platinum or palladium rather than buy commercially produced silver-based paper. At his day job at the Gravure, he advanced the practice of duotone reproduction and developed techniques of tritone and even quadratone imaging that added detail, tonal range and subtlety to an already sophisticated process. He was also an energetic and enthusiastic teacher of his craft, always ready to share his knowledge, techniques and insights with his colleagues in the department. Benson was at Meriden for half a dozen years from the mid-sixties into the early seventies. At that time, he left Meriden to freelance. He was commissioned to photograph the St. Gaudens memorial to Robert Gould Shaw and the 54th Massachusetts Volunteers, which stands on the Beacon Street side of the Boston Common. From these photographs he made duotone negatives[1] for the Eakins Press publication *Lay This Laurel* (see Fig. 4.12), which was printed at Meriden in 1973 under Benson's supervision. Benson also made the tritone negatives for the Museum of Modern Art's monumental four-volume publication of *The Work of Atget,* printed at Meriden starting in 1981 and finishing in 1984.

One employee who joined the camera department during Benson's tenure and who profited from his tutelage was Chien-Fei Chiang. He was the son of Chiang Yee, the author of the *Silent Traveller* books. Chien-Fei was born in China and studied art there before removing to Taiwan when the Communists gained ascendency on the mainland. Through his friendship with Walter Whitehill, Yee became acquainted with Harold. He persuaded Harold to sponsor Chien-Fei's immigration application to come to the U.S. together with his wife and young daughter. This required that Harold guarantee employment to Chien-Fei when he got here. Harold and Parker considered this, and decided that it could and should be done. Between the time that the immigration papers were submitted and the time that the family was approved to enter the U.S. there was an addition to the family, this time a son. Since the son was not provided for on the paperwork, he could not come. He had to stay behind in Taiwan with Chien-Fei's mother-in-law until his paperwork was duly processed and he could join his family in Meriden.

Harold put Chien-Fei to work in the camera room, where he quickly applied his artistic talents to reproductive photography and became a valued and productive worker. Chien-Fei, on his own time, worked in watercolors in a more or less traditional Chinese manner. His work was well received, at first in Meriden and subsequently throughout Connecticut. He remained employed at the Gravure until the time of the consolidation in 1989, at which time he was well enough established to devote himself fully to his own art.

Stephen Stinehour was another with an avid interest in photography. Rocky's eldest son came down to work at Meriden at the time of the merger of ownership in 1977, but he did so by way of Newport, Rhode Island, where he studied offset camera technique with Richard Benson. At Meriden his preferred starting assignment was the camera department. Later he moved out of the shop and took up sales and administrative responsibilities, but he was always keen to undertake photographic books. He successfully solicited work from the Metropolitan Museum of Art, the National Gallery of Art, Princeton University Art Museum and major museums active in the exhibition and publication of photography. Specialist publishers, among them Aperture and Calloway Editions, and trade publishers such as Bulfinch Press were also important sources of work.

The postwar period was a generally prosperous time for America and for the Meriden Gravure Co. It benefitted from increased activity in the sciences and from increased scholarship and public interest and patronage in the arts. Particularly helpful was the establishment of the National Endowment for the Humanities and the National Endowment for the Arts as a part of the Great Society program. Increased activity of non-government organizations such as the American Federation of Arts also helped. Many of the publications printed from the sixties onward were made possible by grants from these programs. One of the ironies of entering the museum market was that instead of printing silverware illustrations for industrial producers, Meriden was now printing the catalogues of the great silver collections in the Yale University Art Gallery, the Wadsworth Atheneum and the Museum of Fine Arts, Boston.

Time in production varied considerably. Occasionally it was only weeks, as in printing the reproduction of the Declaration of Independence in Jefferson's hand, described by John Peckham in *Adventures in Printing*. Usually it was a matter of months from start to finish. Some publications took even longer. *Benjamin Franklin's Philadelphia Printing*, also described by Peckham, was started in 1963 and was not published until 1974. Agnes Mongan's definitive study of nineteenth-century French drawings in the Fogg Art Museum, *David to Corot*, was begun at Meriden in 1969 and completed at the Stinehour Press well after the merger, in 1996. In fairness to Miss Mongan, it must be said that during this time she had teaching and curatorial duties throughout, and for several of those years served as the director of the Museum.

The Yale University Library published *Alchemy and the Occult: A Catalogue of Books and Manuscripts from the Collection of Paul and Mary Mellon*, in four large volumes. The first two volumes, covering the printed books, were published in 1968. They were designed by Joseph Blumenthal. No effort was spared in the production. In order that the illustrations be fully integrated into the text, the collotype illustrations, approximately 160 of them, were scattered throughout the 580 text pages. The collotype sheets were sent down to New York, and the letterpress was struck in at the Spiral Press. Volumes 3 and 4, which catalogued the manuscripts, were not ready for pub-

1719

150 SALLWIGT, Gregorius Anglus, *pseudonym* of Georg von Welling, 1652–1727

OPUS MAGO-CABALISTICUM ET THEOLOGICUM. Frankfurt-am-Main 1719

יהוה OPUS MAGO-CABALISTICUM ET THEOLOGICUM. Vom Uhrsprung und Erzeugung Des Saltzes, Dessen Natur und Eigenschafft, Wie auch Dessen Nutz und Gebrauch. Da denn zugleich die Erzeugung aller Metallen und Mi-neralien, und aller andern Salien aus dem Grunde der Natur bewiesen wird; Auch viel Theosophica, nach Gelegenheit der Materien / mit untergemischt werden. Deßgleichen auch weitläufftig discuriret wird von denen uns unsichtbaren Creaturen / in denen uns sichtbahren und greifflichen Elementen / wie auch von dem Paradiese und dessen Loco, welches alles vorgestellet wird durch das Systema Magicum Universi; dadurch der Wahrheit-liebende zu den allerhöchsten und heiligsten Geheim-nüssen geleitet und geführet wird. Alles auffgesetzt und zusammen getragen von einem embsigen Liebhaber der ewigen Wahrheit / dessen Nahmen GREGORIUS ANGLUS SALLWIGT überkommen ANNO MDCCVIII. Und Franckfurth am Mayn gedruckt bey Anton Heinscheidt / 1719.

COLLATION: 2°: a-b² A-U² ($2 signed); pp. [8] 1-80.

CONTENTS: a1: title. a1ᵛ: □. a2: preface, dated at end 'Franckfurth am Mayn / den 28. | Mertz 1719.' and signed 'S. R.' A1 (p. 1): text.

ILLUSTRATIONS: Ten plates, of which four are folding, engravings, all except one hand-colored, of hermetic diagrams. Woodcut symbols in margins.

DECORATIONS: Type-ornament headpieces. One woodcut tailpiece, vase of flowers.

TYPOGRAPHY: M2ʳ – 66 lines, 274 (284) x 148 mm., 84 (11.96) gothic and roman. Catch-words throughout. Running titles: (A-E) CAP. I. | DE ORIGINE SALIS COMMUNIS. (F-H) CAP. II. | DE NATURA SALIUM. (H-M) CAP. III. Von dem Nutz und Gebrauch des Saltzes.

520

יהוה

OPUS MAGO-CABALISTICUM ET THEOLOGICUM.

Vom Uhrsprung und Erzeugung Des Saltzes, Dessen Natur und Eigenschafft, Wie auch Dessen Nutz und Gebrauch.

Da denn zugleich die Erzeugung aller Metallen und Mineralien, und aller andern Salien aus dem Grunde der Natur bewiesen wird;

Auch viel Theosophica, nach Gelegenheit der Materien/ mit untergemischt werden.

Deßgleichen auch weitläufftig discuriret wird von denen uns unsichtbaren Creaturen, in denen uns sichtbahren und greifflichen Elementen, wie auch von dem Paradiese und dessen Loco, welches alles vorgestellet wird durch das Systema Magicum Universi, ...

Alles auffgesetzt und zusammen getragen von einem embsigen Liebhaber der ewigen Wahrheit, dessen Nahmen

GREGORIUS ANGLUS SALLWIGT

überkommen

ANNO MDCCVIII.

Und Franckfurth am Mayn gedruckt bey Anton Heinscheidt, 1719.

NO. 150

FIG. 4.5. *Alchemy and the Occult From the Collection of Paul and Mary Mellon*, Volume 2, 1968, designed by Joseph Blumenthal. The illustrations were printed in collotype at Meriden, and the text was printed letterpress at the Spiral Press. It was published by the Yale University Library.

lication until the mid-seventies. By that time Meriden was no longer doing collotype and Blumenthal had disbanded the Spiral Press. Blumenthal did the design; Stinehour did the typesetting. The presswork, text and illustrations were printed in offset at Meriden in 1977.

Another decades-long project was the Houghton *Shah Nameh*, a facsimile of a magnificent sixteenth-century Persian manuscript, one of the most important of its kind in western hands. Work on the black-and-white collotype illustrations began in the early sixties and was completed before the collotype department was closed down. It was eventually published in 1981 by Harvard University Press after the accompanying scholarly text by Stuart Cary Welch and Martin Bernard Dickson was completed.

Whatever the production time of a job, on-time delivery was always a crucial consideration. Publishers needed their deliveries on a timely basis to take advantage of the Christmas selling season or to coordinate with advertising, promotion and marketing efforts. Museums needed their catalogues for exhibition openings. Failure to have copies at the opening put a damper on the festivities and resulted in embarrassment for all concerned. Sales were lost, only some of which could be made up later. The Gravure would move heaven and earth to make an on-time delivery.

Glick remembers a sixty-four-page paperbound catalogue on the work of Willem de Kooning, produced in 1965 for Charles Scott Chetham of the Smith College Museum of Art. It was printed and bound, and advance copies were delivered to the Museum. It was only when the Museum staff went to hang the pictures that they realized that their layout instructions to Meriden were incorrect and the cover illustration, printed in color, was upside down. The transparency used as copy had no signature, and the abstract character of the image did not suggest a top or a bottom. Meriden went back to press and reprinted the covers. Mueller Trade Bindery pulled the old covers off and stitched on the new ones. The same image appeared in black and white in the interior; miraculously, this image had been printed right side up. Ten minutes before the opening reception Glick, suitably attired for the event in his dinner jacket, starched shirt and well-shined shoes, was at the Museum loading dock to deliver the job.

The University Presses

Another development helpful to Meriden was the insistence of colleges and universities that their faculty members do publishable research as well as teaching. The "publish or perish" syndrome spurred the growth of existing university presses and prompted the establishment of new ones. The university presses were an important source of business for Meriden. The Gravure took an active interest in the American Association of University Presses. It frequently exhibited at the annual convention. The exhibition of the best-designed and best-produced books from AAUP members invariably included books printed at Meriden.

Of particular importance to Meriden was the Princeton University Press. University presses tended somewhat to specialization in certain academic disciplines. Princeton was the premier publisher of scholarly books on art history and classical archaeology. One of its editors, Harriet Anderson, was particularly knowledgeable in this area. The practice of art historical scholarship in America was significantly transformed by the wave of European, mainly German, art historians fleeing the Continent in the thirties. Many

Fig. 4.6. Illustration printed in offset for *Michelangelo: Sculptor, Painter, Architect,* Charles De Tolnay's one-volume distillation of his five-volume study of the artist. Princeton University Press, 1975.

found asylum in the U.S. and secured teaching positions in American colleges and universities. They brought with them rigorous standards of scholarship and a long and intimate exposure to Europe's artistic heritage. Princeton University Press was the publisher of many of their magisterial works.

Erwin Panofsky's *Dürer* (1943), Walter Friedlander's *Caravaggio Studies* (1955) and Richard Krautheimer's *Lorenzo Ghiberti* (1956) were among the major contributions to the literature published by Princeton with illustrations printed at Meriden. Charles De Tolnay's *Michelangelo, The Medici Chapel* (1948) included the author's expression of thanks to Harold for his "careful supervision of the printing of the collotype reproductions," a note which was carried forward to the 1970 reprinting, in which the plates this time were printed in offset. *The Medici Chapel* was one of five volumes on Michelangelo written by De Tolnay and published by Princeton between 1943 and 1960. Meriden also printed illustrations for a one-volume distillation of the author's research, *Mi-*

chelangelo: Sculptor, Painter, Architect (1975). H. W. Janson's *Donatello* was published in 1957 in two volumes, the first volume being 512 plates printed in collotype by Meriden, and the second being the text volume printed at Princeton. The success of this publication created a demand for a one-volume edition, in which the illustrations were whittled down to 316 on 128 pages printed in 300-line offset at Meriden in 1963.

The College Art Association, whose members were college-level teachers and researchers in the fine arts, was publisher of the academic quarterly *Art Bulletin*. Harriet Anderson served as managing editor for many years until Jean Lilly succeeded her in that position. *Art Bulletin* was prepared for publication at the Princeton University Press. The text was printed at Princeton's printing facility, and the illustration sections were printed at Meriden and shipped to Princeton for binding.

Also on the schedule for any trip to Princeton was the American School of Classical Studies at Athens, which was a part of the scholarly environment of Princeton but with a somewhat autonomous governance. Meriden provided the illustrations for their book-length monographs and their quarterly publication, *Hesperia*. The text printing for these publications, as for the College Art Association publications, was done letterpress at Princeton.

The presses of the other Ivy League universities were also sources of work for Meriden. Yale and Harvard both had robust programs, although Harvard was not as active in the visual arts as some of its sister institutions. Outside of the Northeast the presses of the University of Oklahoma and the University of Texas, both funded by oil revenue, had strong programs in local and regional history and culture including architecture, decorative arts and folk art. The University of California at Berkeley, Louisiana State University and the University Press of Virginia were frequent customers of Meriden, as were the university presses at Chicago, Michigan and Wisconsin. Locally, the Wesleyan University Press in Middletown, eight miles over the hill from Meriden, had a very active publishing program for an institution that was largely devoted to undergraduate liberal arts instruction.

The Papers of the Founding Fathers

One of the great scholarly enterprises of the postwar period was the publication of the papers of the founding fathers. The university presses were deeply involved in this. The first of these multi-volume, multi-year publications was *The Papers of Thomas Jefferson*, meticulously edited by Ju-

Plate 11.—Casa del Gobernador, Uxmal

FIG. 4.7. Detail of foldout plate, one of 80 illustrations copied from the original 1843 edition of *Incidents of Travel in Yucatan* by John Lloyd Stevens for the 1962 edition published by the University of Oklahoma Press.

lian Boyd, designed by P. J. Conkwright and published by the Princeton University Press. *The Adams Papers* were published by the Harvard University Press. Boyd's associate editor, Lyman Butterfield, was recruited to Boston to lead the editing team working at the Massachusetts Historical Society, where the bulk of the Adams family papers were housed. Yale was the publisher of *The Papers of Benjamin Franklin*, Columbia did *The Papers of Alexander Hamilton*, the University of Chicago Press did James Madison, and South Carolina did Henry Laurens. These scholarly projects were not limited to the founding fathers. *The Papers of Woodrow Wilson*, edited by a team led by Professor Arthur Link, was also a Princeton U. P. publication. The papers of other American statesmen were edited and published, frequently by the university press of their home state.

All of these editions were long on text but flavored with a modest number of illustrations. When there were illustration inserts to be done, Meriden usually did them. The notable exception to this were two volumes in the Adams Papers series, *Portraits of John and Abigail Adams* by Andrew Oliver, published in 1967, and *Portraits of John Quincy Adams and His Wife*, also by

Fig. 4.8. Jacket for *New York Landmarks*, published under the auspices of the Municipal Art Society of New York by the Wesleyan University Press, 1963.

Oliver, published in 1970. These titles, each with more than one hundred illustrations scattered throughout the text, were printed entirely by Meriden.

One of the guiding figures in this great enterprise was the historian Philip May Hamer, who served as Executive Director of the National Historical Publications Commission. This government body coordinated and encouraged the work of the scholars and publishers of the various titles in the program. In 1960 Hamer was honored with a dinner and a publication that included Meriden's contribution of seven important historical documents in collotype facsimile. Harold was thanked for his "devotion to the high standards of quality in printing . . . matched by his indefatigable labors and generosity in support of useful enterprises in the world of scholarship."[2]

Henry Laurens as President of the Continental Congress, by John Singleton Copley. *National Portrait Gallery, Smithsonian Institution, Washington, D.C.*

The Papers of
Henry Laurens

Volume One: Sept. 11, 1746–Oct. 31, 1755

Philip M. Hamer, Editor
George C. Rogers, Jr., Associate Editor
Maude E. Lyles, Editorial Assistant

Published for the South Carolina Historical Society by the
UNIVERSITY OF SOUTH CAROLINA PRESS
COLUMBIA, S.C.

FIG. 4.9. Color frontispiece and title page for the *Papers of Henry Laurens*, 1968.

The Trade Publishers

The major trade publishers occasionally called upon Meriden to do an especially demanding or important job, but this was not a mainstay of Meriden's business. Wilmarth Lewis had enough clout to insist that Meriden print the illustrations in his autobiography, *One Man's Education*, published by Knopf in 1968. But in general, the tight production budgets and the length of the press runs precluded this. However, some small trade publishers with high aspirations and great determination to do publishing better than is ordinarily thought necessary did have work done at Meriden.

David Godine was one of Ray Nash's most ambitious and enterprising students and went on to found his eponymous publishing firm in Boston. Originally he attempted to be both a printer and a publisher, but subse-

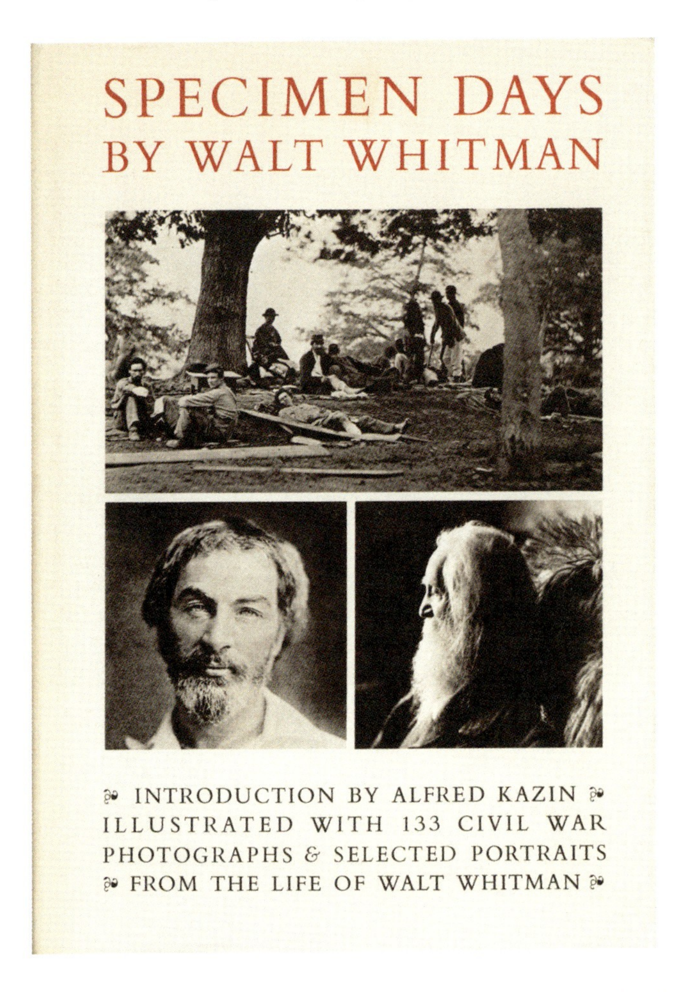

FIG. 4.10. Jacket for the Godine Press edition of *Specimen Days*, 1971. Edited and designed by Lance Hidy.

quently gave up the printing to concentrate on publishing. Godine had wide-ranging interests, which included literature, criticism, poetry, photography and cultural studies. But he always found room on his list for worthy additions to the already vast literature of books on books. He was publisher of Joseph Blumenthal's *Art of the Printed Book* and *The Printed Book in America*. The illustrations for each of these titles were printed at Meriden.

An important early undertaking of David R. Godine, Publisher was *Walt Whitman's Specimen Days* (1971). This was a reprinting of a great collection of autobiographical notes and observations made by Whitman. It had been printed in several editions in the nineteenth century, but had long been out of print. The Godine edition was largely the effort of Lance Hidy, who researched, edited and designed it. This edition was greatly enhanced by Hidy's

FIG. 4.11. Jacket for George Tice, *Urban Romantic.* Lettering by G. G. Laurens. David R. Godine, Publisher, 1982.

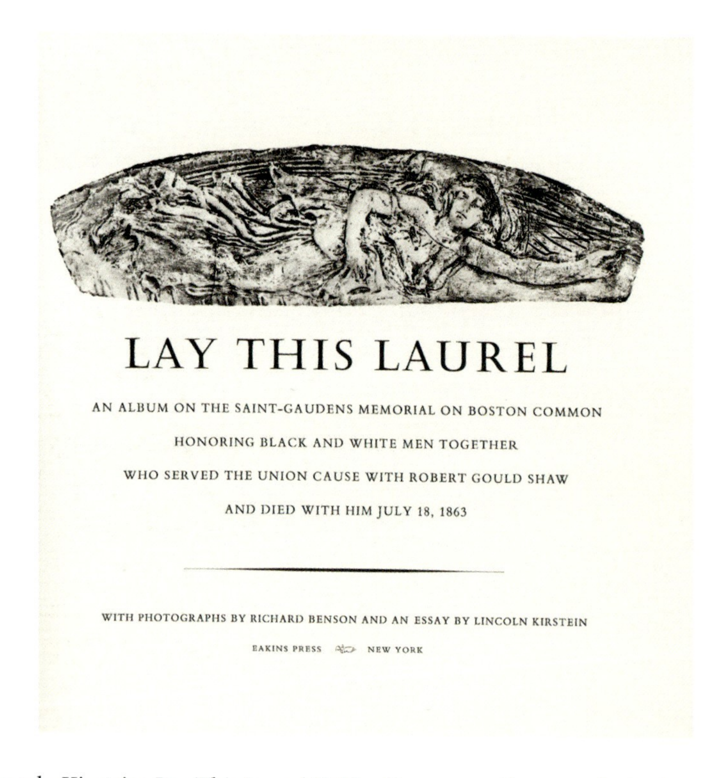

FIG. 4.12. Lincoln Kirstein, *Lay This Laurel,* Eakins Press, 1973. Typography by Freeman Keith.

inclusion of 133 vintage photographs. Some were life portraits of Whitman and others, taken by Matthew Brady and his contemporaries, recording the Civil War environment that Whitman experienced. The text was printed at Meriden from type set at Stinehour, with the photos done in duotone.

A number of Godine's photography books were printed at Meriden. George Tice's *Urban Romantic* (1982) was done at Meriden, with Tice very much on the scene when the job was on press. Tice's *Fields of Peace* was done in 1998 in Lunenburg. Kipton Kumler's very formalist photography was represented in *Plant Leaves* (1978) and, at the other end of the photography spectrum, Barbara Norfleet's documentary *Champion Pig: Great Moments in Everyday Life* appeared the following year.

Among other noteworthy publications done at "the Grav," as Godine liked to call it, was *Boston: Distinguished Buildings and Sites.* The reproduction of Rudolph Ruzicka's color woodcuts, originally done as the New Year's greetings of the Merrymount Press, were here gathered together with brief commentaries by Walter Whitehill. It is among the books described in detail in *Adventures in Printing.*

Leslie Katz founded the Eakins Press in 1966. His interests were widespread, but centered on the American cultural scene and the visual and performing arts. He was the publisher of the facsimile of the first edition of *Leaves of Grass* and of *Lay This Laurel*, both mentioned previously. Other publications that he brought to Meriden were Hilton Kramer's *Sculpture of Gaston Lachaise* (1967) and, in the same year, a reduced-size reprinting of Louis Sullivan's *A System of Architectural Ornament*, originally published in 1924.

Katz was a friend and close associate of Lincoln Kirstein and through him did a number of publications on George Balanchine's New York City Ballet. One such publication printed at Meriden was photos of the ballet *Union Jack* taken by Richard Benson and Martha Swope.

Barre Publishers was an offshoot of a small local paper, the *Barre Gazette*, published by Alden Johnson in Barre, Massachusetts. Johnson, like Katz and Godine, was a man of wide interests, but his greatest concentration was in American history and popular culture. His publication of *Massachusetts Silver in the Frank L. and Louise C. Harrington Collection* was printed by Meriden in 1965. Wendell Garrett's *Thomas Jefferson Redivivus*, with photographs by Joseph Farber and typographic design by Klaus Gemming, went through Meriden in 1971.

Other important publications from Barre were the *Views of Ancient Monuments in Central Chiapas & Yucatan* by Frederick Catherwood, which is discussed in *Adventures in Printing*, and *Moby Dick: The Passion of Ahab*,

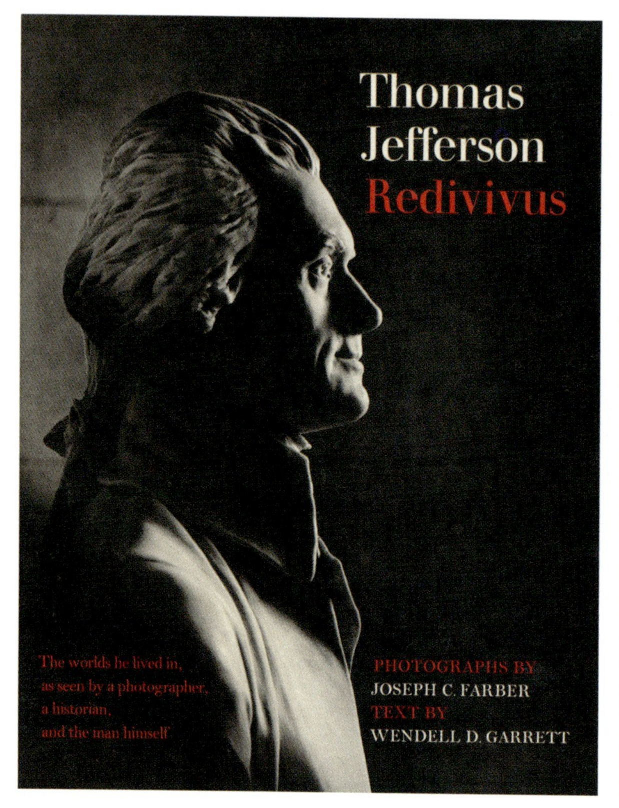

FIG. 4.13. Wendell D. Garrett, *Thomas Jefferson Redivivus,* Barre Publishers, 1971. Design and production by Klaus Gemming. FIG. 4.14. *Artist's Proof,* Volume VIII, 1968. The Pratt Graphics Center in association with Barre Publishers. Design and production by Bert Waggott.

mentioned elsewhere in this book. A major undertaking for Barre was the publication in facsimile of *The Atlantic Neptune,* an atlas of nautical charts of the Atlantic coast of North America published between 1774 and 1781 by the British Admiralty. The fifty-nine charts were issued in four large 24 x 36-inch portfolios. The engraved key plates and the color borders were printed by offset at Meriden, and the rest of the color was done by hand at the Berrien Studio.

Barre also took on *Artist's Proof, The Annual of Prints and Printmaking* as co-publisher (with the Pratt Graphics Center) and as distributor. *Artist's Proof* was a significant influence in the renewal of interest in printmaking by artists, dealers, museums and the art-conscious public in the decades following the Second World War. It was founded by Fritz Eichenberg, Director of the PGC, a charismatic teacher and ardent exponent of graphic art in any medium and in any style, from traditional to the most avant-garde. Harold greatly admired Eichenberg for his commitment and had been printing the publication at cost or at what the budget would allow, whichever was less. But it needed better financing and better marketing, and through Harold's influence Barre agreed to the arrangement.

Johnson also established the Imprint Society, a subscription book club modeled somewhat along the lines of the Limited Editions Club. But where the LEC published the recognized high points of literature, the Imprint Society sought out the neglected classics in history and literature. The editorial board assembled by Johnson included Walter Whitehill, Thomas R. Adams, Lawrence Thompson, James F. Beard (the editor, not the food guru) and David McCord. Like the LEC, the Imprint Society went to great lengths to work with the most prestigious designers and printers in America and Europe. Among the books printed at Meriden were *Sketches in North America and the Oregon Territory* by Capt. H. Warre, the commander of a British Army Expedition to those parts in 1845–46, and *The Complete Almanacks of Poor Richard* by Benjamin Franklin. The illustrations for a reprint of the 1840 edition of N. P. Willis, *American Scenery*, with engravings of W. H. Bartlett, was done in association with the Anthoensen Press; the illustrations for a new English translation of a Renaissance Italian manuscript of *The Labors of Hercules* was a collaboration with the Stamperia Valdonega of Verona, Italy. The reproductions of the suite of Otto Dix etchings, *Das Kreig*, for *Bellum: Two Statements on the Nature of War*, *The Boston Gazette 1774* and *American Broadsides, Sixty Facsimiles dated 1680 to 1800 reproduced from originals in the American Antiquarian Society* were all done in association with David Godine.

David R. Godine, Publisher continues to thrive more than forty years after its founding. Although Leslie Katz died in 1997, the Eakins Press lives on as a nonprofit foundation and continues to publish. Barre Publishers and the Imprint Society, on the other hand, did not long survive the death of Alden Johnson in 1972.

The Bicentennial Year

Meriden had a particularly busy year in 1976, the year of the National Bicentennial. The surge started in 1975, marking events in Massachusetts, peaked in 1976 and did not subside until 1977. Towns and cities, states, the federal government, libraries, colleges and universities and foreign nations marked the occasion with exhibitions and publications, many of which were produced at Meriden. Among these were *A Rising People: The Founding of the United States, 1765 to 1789*, published by a consortium of Philadelphia institutions: the American Philosophical Society, the Historical Society of Pennsylvania and the Library Company of Philadelphia; *Wellsprings of A Nation: America Before 1801*, published by the American Antiquarian Society; and *Thirteen Colonial Americana*, edited by Edward Connery Lathem published by the Association of Research Libraries. *Early Vermont Broadsides* was published by the University of New England Press; *Rebellion and Reconciliation: Satirical Prints on the Revolution at Williamsburg* by Joan Dolmetsch was published by the Colonial Williamsburg Foundation.

The National Gallery of Art gave Jefferson his due in two lengthy books, *The Eye of Thomas Jefferson* and *Jefferson and the Arts: an Extended View.* Old Sturbridge Village issued *The Village and the Nation* as its contribution to the observance. American art of the colonial and federal period was celebrated at the MFA in *Paul Revere's Boston, 1735–1818*; *The Classical Spirit in American Portraiture* was published by Brown University and *Where Liberty Dwells: 19th Century Art by the American People, Works of Art from the Collection of Mr. and Mrs. Peter Tillou* was exhibited at museums in Utica and Buffalo, New York, Milwaukee and Philadelphia.

The National Portrait Gallery under the leadership of Director Marvin Sadik organized an exhibition, *Christian Gullager, Portrait Painter to Federal America*, with accompanying catalogue designed by Leonard Baskin, typeset at Stinehour and printed at Meriden. Gullager was born in Denmark and trained as an artist there. He came to America as a young man, and he enjoyed much success in his calling in the young republic. The exhibition and catalogue were largely financed by the Danish government as its contribution to the celebration of the American Bicentennial.

The Caxton Club of Chicago published *Jefferson's A Summary View of the Rights of British America*, an argument for independence directed at Virginians. *The Suffolk Resolves, 1774*, edited by James Mooney and published by the Society of Printers, Boston, expressed the stiff resistance to British economic sanctions against Massachusetts, while *Harvard Divided* by Linda Ayres, published by the University, was a reminder that there was considerable Tory sympathy in the Harvard community at the time.

Collaborators

Throughout this time the Gravure continued to limit itself to illustration printing, working in conjunction with other graphic arts firms that did design, typesetting, letterpress printing and binding. Among the most nota-

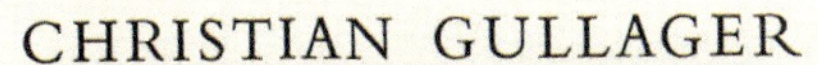

Fig. 4.15. *Christian Gullager: Portrait Painter to Federal America*, one of the many Bicentennial publications. Design by Leonard Baskin.

ble of these printers in New York were Joseph Blumenthal's Spiral Press and Clarke & Way, a partnership of Bert Clarke and David Way.

Blumenthal wrote in *Typographic Years: A Printer's Journey Through a Half Century, 1925–1975*, "If I were asked by a harassed St. Peter outside the Pearly Gates to show only one printed book on which I should be judged, I would hope to have with me the weighty volume *Chinese Calligraphy and Paintings in the Collection of John M. Crawford, Jr.*" [3] He continues, "I turned to Harold Hugo and John Peckham at Meriden, friends and colleagues in pursuit of the

FIG. 4.16. Mynah Bird and Bamboo, Plate 30, *Chinese Calligraphy and Painting in the Collection of John M. Crawford, Jr.* Printed in collotype. Pierpont Morgan Library, 1962.

maximum integration between type and plates." This volume, published in 1962, was a collaboration both with Meriden, which printed fifty-one black-and-white collotype illustrations, and with Arthur Jaffé, New York, which printed three foldout color collotype plates. Meriden participated in other publications that Blumenthal singled out as among his proudest achievements. These included collotype work for the facsimile edition of Thackeray's manuscript of *The Rose and the Ring* (Pierpont Morgan Library, 1947), *The Italian Influence on American Literature* by C. Waller Barrett (The Grolier Club, 1962), and *To Russia With Frost* by Frederick B. Adams, Jr. (The Club of Odd Volumes, 1963). Offset illustrations were provided for *Poetry in Crystal* (Steuben Glass, 1962), *The Grolier Club 1884–1967* by John T. Winterich (The Grolier Club, 1967), *Eighteenth-Century Studies in Honor of Donald F. Hyde*, edited by W. H. Bond (The Grolier Club, 1970) and *Robert Frost and His Printers*, written and designed by Blumenthal, text printed letterpress at A. Colish, Mount Vernon (W. Thomas Taylor, Publisher, 1985).

The Anthoensen Press; Connecticut Printers in Hartford; the presses of Yale and Princeton universities; and the Stinehour Press of Lunenburg, Vermont, were also important both as providers and as customers. Thomas Todd in Boston was both a competitor and a collaborator. A considerable amount of work came from independent designers. Malcolm Grear Designers in Providence; Klaus Gemming, Nathan Garland, Howard Gralla and Greer Allen, all from New Haven; and Bradbury Thompson of Greenwich, Connecticut, were important sources of work. These designers had various favorite sources of type composition. Stinehour was one of them. Finn Typographic Service in Riverside, Connecticut; Michael Bixler Letterfoundry in Skaneateles, New York; and Mackenzie & Harris in San Francisco were also highly regarded and frequently used. Carl Zahn of the Museum of Fine Arts, Boston, and Peter Oldenburg, the chief designer for the Metropolitan Museum of Art in New York, favored Meriden with much of their work. Richard Hendel brought work to Meriden from the University of Massachusetts Press and the University of North Carolina Press, as well as various freelance assignments.

All of these firms including the Gravure were dependent upon one another for those services they could not themselves provide. Each contractor necessarily held his suppliers to the same high standards of craftsmanship and the same obligation to on-time delivery as prevailed in his own shop.

Further Adventures in Printing

In addition to the books that he talked about in *Adventures in Printing*, John Peckham recorded other noteworthy adventures. He gave a talk to the Philobiblon Club of Philadelphia in 1985 and discussed a number of important books that Meriden had produced for Philadelphia customers. John had a very acute memory for detail, and his commentary on three of the publications is here put into print with some of the informality of his speaking style edited out.[4]

J. Eric Thompson, *A Commentary on The Dresden Codex*. Philadelphia: The American Philosophical Society, 1972.

The text of this book was designed and printed letterpress in 1972 at the Stinehour Press. Meriden produced the illustrations. The story behind these plates is an extraordinary one.

In all the world there exist only three hieroglyphic books made by the Mayan Indians predating the discovery of America by Christopher Columbus. The Spanish discoverers have described whole libraries of these books, which they systematically destroyed in their goal of Christianizing the heathen Indians. These books, called codices, were made of bark fibers compacted in the same way that Egyptians made papyrus, and with the surfaces coated with a white sizing of lime, on which the hieroglyphics were drawn. Each codex consisted of panels hinged to each other to create an accordion format, arranged so that facing pages could be read when each two-page spread was opened. The panels were typically about 3-½ inches wide by 8 inches high, and a book when opened would constitute a strip 11-½ feet long.

The three surviving codices have been named for the cities where they now repose: Madrid, Paris and Dresden. The Dresden Codex is presumed to have been sent in 1519 by the conquistador Cortés to the Holy Roman Emperor, Charles V, in Vienna and was purchased in 1739 by the Saxon Royal Library of Dresden. During World War II it suffered water damage in the Allied fire-bombing of the city. Meriden's reproduction was made not from the damaged original but from a very fine chromolithographed reproduction made in Germany in 1892, and artists produced overlays of the colors, which we copied and printed in register. Several mistakes by earlier scholars were corrected, including the sequence of pages. The original codex had broken at two places, and Thompson authoritatively demonstrated how the sections followed each other. This corrected previous studies that had assumed a sequence in which an entire section was flopped.

Joseph Ewan, *William Bartram: Botanical and Zoological Drawings 1756–1788.* Philadelphia: The American Philosophical Society, 1968.

The page size of 11-⅛ x 15-⅛ inches was close to the longest we could run, four pages at one time on our presses. The text was composed and run by letterpress at the Stinehour Press. William Bartram had made his beautiful drawings during three trips in the southern states, Florida, Georgia, Alabama and North and South Carolina. His patron for this trip was an Englishman, Dr. John Fothergill. These drawings, bound together in an album, were eventually acquired by the Botanical Library of the British Museum. In 1966 the APS decided to publish the book and sent Harold to London to supervise the photography of the plates. When he returned to Meriden, he had proofs made of a few typical plates and studied them for quality. A puzzling phenomenon occurred: though the images were faithful to the original drawings in color and detail, they appeared dull and flat compared to the originals. Harold then realized that the originals had been pasted in an album the pages of which were a rather dark gray color. Our proofs had been pulled on a very nice cream-colored paper. Harold had gray background approximately like that in the album added around the drawing images, and this immediately retrieved the brilliance and drama of the originals.

Moby Dick: The Passion of Ahab. Twenty-six Lithographs by Benton Spruance with text by Lawrence Thompson. Barre, Massachusetts: Barre Publishers, 1968.

It all started when Larry Thompson, biographer of Robert Frost and professor at Princeton, asked me to look at a couple of lithographs by Benton Spruance, the well-known artist-lithographer. Spruance's studio was in a tiny eighteenth-century building, formerly a store with a big bay window on a street in Germantown, Pennsylvania. In the front room was the biggest lithographic press I have ever seen, on which Spruance printed from the biggest stone I had ever heard of—about 28 x 40 inches, and three inches thick. He had a cart with rollers at the height of the press bed on which he slid the stone on and off the press. Ben had worked out a system whereby he could, after printing a color, chemically neutralize the image on the stone so that it would not take ink, but the image would remain visible. He could then draw the image for the next color in register on the previous image. He would repeat this process until he had printed all the colors he wanted to run.

Thompson had written a book, *Melville's Quarrel with God.* Spruance had read it and become enthralled with the symbolism and images in Moby Dick, and proposed to do a series of lithographs derived from the book. The two prints I was shown were the only ones done at that time, and I realized that an extraordinary opportunity pre-

FIG. 4.17. Prospectus for Barre Publishers' *Moby Dick: The Passion of Ahab,* 1968.

> sented itself. If Spruance would provide proofs of each of his colors for every print, we could copy those proofs and run the colors one at a time, matching his inks, instead of copying from the finished prints and running ordinary four-color process. Although the portfolio pages we printed were 18 by 23 inches, about half the size of Spruance's originals, they faithfully matched the artist's colors. Upon publication, the Philadelphia Museum of Art mounted an exhibition of our reproductions side by side with Spruance's original prints, than which nothing could have been more demanding. The exhibition was also to honor the great collector and benefactor, Lessing Rosenwald, who acquired the key set of Spruance's prints (the artist's edition was seven copies) and also the artist's progressive proofs.

What John in his modesty did not say was that this project was entirely his project and that Harold had no direct involvement in it. Further, that he took great satisfaction in once again working with Larry Thompson, who had been his thesis advisor at Princeton almost thirty years earlier when Thompson had been Curator of Special Collections.

Another difficult but memorable job produced at Meriden was the catalogue *American Painting and Sculpture*, produced under the auspices of the U.S. Information Agency. This was a part of the 1959 exhibition of American culture in Moscow at which Vice-President Nixon famously debated Chairman Khrushchev on the relative merits of free enterprise versus state socialism at the display of American consumer goods. The production of the art catalogue ran a troubled course, as was explained to Harold and John by Caroline Rollins, who was involved in organizing the exhibition and producing the catalogue. The text was written in English and translated into Russian by émigré aristocrats living in Paris. After this work was done, it was discovered that their Russian was an old-fashioned courtly Russian that was no longer used in the present-day peasant's and worker's paradise. The text had to be completely retranslated into contemporary Russian, creating delays that the production schedule could ill afford. The other problem was that the art selected for exhibit was entirely twentieth-century art, including works of the Ash Can School from the twenties and social protest art of the thirties. This did not present America as the land of milk and honey. McCarthyism had peaked by this time, but it had not entirely disappeared, and some politicians put up a howl that led to further delays. The organizers managed to negotiate these shoals, and the publication, 100 pages of black and white plus six color plates, was printed, bound and shipped off to Moscow in time for the opening, but only just barely. The catalogue was eagerly received by visitors to the exhibition, but ironically, the content was as offensive to the Soviet cultural bureaucracy as it was to some American politicians. However in the USSR, the bureaucrats prevailed, and the undistributed copies were soon confiscated and destroyed.[5]

The largest single publication that was done in the offset era was the "Schuster/Carpenter." Its full name was *Materials for the Study of Social Symbolism in Ancient & Tribal Art: A Record of Tradition & Continuity.* It was based on the research and writing of Carl Schuster (1904–1969), and was edited and written by his colleague, supporter and literary executor, Edmund Carpenter (1922–2011). Schuster was a dedicated field anthropologist, the master of some thirty languages, whose research led him to all parts of the inhabited world. His interests ranged from ancient societies to contemporary ones. It was his conviction that correspondences and connections existed between cultures widely separated by space and time. Anthropologists and ethnologists who were narrow specialists dismissed this approach. As Carpenter writes, "the notion that incised stones in Spain might relate to similar stones in Patagonia seemed absurd to specialists."[6] Carpenter took

1. БУМ-ТАУН, 1928

ТОМАС ХАРТ БЕНТОН

Бентон родился в 1899 году в штате Миссури и был назван в честь своего известного дяди, который был сенатором от этого штата. По окончании высшего образования, Бентон поступил в чикагский Институт искусств. Он провел некоторое время в Париже, но обрел свой талант художника после возвращения в Соединенные Штаты и знакомства с Американским Западом — нескончаемыми степями, равнинами и другими сюжетами для художественного изображения.

Кроме работы у мольберта, художник создал много картин с изображением жизни в его родном и других маленьких и больших городах, рисовал отдельные типы населения, изобразил на полотне легенды родных мест и занимался также росписью стен в целом ряде общественных зданий. В настоящий момент он работает над росписью стен в здании Библиотеки имени Трюмэна в г. Индепенденс, Миссури.

FIG. 4.18. *American Painting and Sculpture.* The U.S. Information Agency for the American Embassy, Moscow, 1959. The page shows Thomas Hart Benton, *Boomtown*, 1928, from the collection of the Memorial Art Gallery, University of Rochester, Rochester, N.Y.

on the task of setting forth Schuster's theories in a comprehensive fashion and getting them into print. He established the well-funded Rock Foundation to publish the work and distribute it free of charge to important institutions of higher learning worldwide. In doing so he avoided the academic publishers who had spurned Schuster's manuscript submissions.

Social Symbolism in Ancient & Tribal Art was published in a generous 11 x 14 format, which could only be run four pages up on Meriden's presses. It had two colors throughout with additional colors where required. It had 7000 illustrations. Its almost 3200 pages, when bound and slipcased, weighed about seventy-five pounds. The work was divided into three "volumes" correspond-

FIG. 4.19. The complete set of Schuster-Carpenter, *Social Symbolism in Ancient & Tribal Art*, with the "Shorter Schuster" perched on top.

ing to the three major divisions of the work. The volumes were further broken down into books, which were bound individually in hard cover, with four books making up Volume 1, five for Volume 2 and three for Volume 3. Each volume had its own slipcase. In this grand enterprise the only feature that was in any way limited was the edition, which was planned for 600 copies but which probably yielded some copies beyond that. Work was begun at Meriden in 1983. Volume I is dated 1986 and Volumes 2 and 3 are dated 1988, but distribution had to wait until all volumes were completed in 1991.

"Shorter Schuster," *Patterns That Connect: Social Symbolism in Ancient & Tribal Art*, was published in 1996. This was a 320-page, one-volume summary by Carpenter of the larger work, with a mere one thousand illustrations. It was printed at the Stinehour Press and published by Harry N. Abrams, Inc.

In general, scientific publications did not have the glamour that was associated with fine arts publications. There were no swish museum openings at which a publication made its debut before a fashionable crowd enjoying first-class food and drink as well as stimulating art. The scientific publications, many of them periodicals, simply came along without ceremony, each one in its own discipline, and each adding further to the body of knowledge in that discipline.

Five

Running the Show

The Meriden Gravure Company as founded by J. F. Allen was run along rather authoritarian and paternalistic lines, as was the case with most old-line Yankee industrial operations. This was continued, but with some relaxation over time, by Parker and Harold. The boss hired you and fired you and could do so on whim and without cause or recourse. Relations with employees were conducted on a direct, face-to-face basis. There was no Employee Handbook codifying and detailing the rights and responsibilities of the employees. These were established by usage and precedent, and were unwritten. Neither was there any "table of organization" or document diagramming the chain of command. Everyone simply knew to whom and for whom they were responsible. There was no "Human Resources" Department, no bureaucracy and very little paperwork. If an employee was caught in an act of dishonesty or in a state of intoxication while on the job, he was summarily dismissed.

Because the Gravure was a leader in its field, it was particularly concerned that experienced, skilled employees should not leave the company and seek employment with a would-be competitor. To do so was considered disloyal. Those who left voluntarily, no matter how talented they may have been, had a very slim chance of being hired back if they had a change of heart and wished to return.

This posture toward competitors did not apply to other local printers serving different markets than the Gravure, and who were not considered competitors. Miller-Johnson was a Meriden firm doing commercial work for local and regional customers, and its owners were socially friendly with Parker and Harold. If Miller-Johnson was in a pinch, the Gravure was willing to lend it paper or ink or equipment, and this cooperation was reciprocal. It was pretty well understood among the managements of the local printers that you did not raid one another's help.

In return for the dedication of its employees, the Gravure provided a level of job security unheard of in the present day. This commitment was most evident during the Great Depression when employees were kept on to do maintenance work when there was no production work at hand. The response to lack of work was for management to get busy and drum up some business. To lay off or downsize a skilled workforce was the option of last resort. In the late 1950s an effort was made to unionize the shop. Harold and Parker strenuously resisted unionization. At this time the shop was busy, and there were no lay-offs and no short hours. On the contrary, there was considerable overtime work, as it was the practice at Meriden to work a half-day on Saturday and the eight-hour workday was sometimes stretched to ten during peak production periods. Most of the employees felt reasonably well-treated and felt that they had more to lose than to gain by becoming a closed shop. When it was put to a vote, the union was voted down, and unionization was never again an issue.

A practical man, Harold was not inclined to think in terms of abstract principles and ideologies. He was not interested in bringing to Meriden the lofty, sweeping theoretical approaches that had become fashionable at the nation's elite business schools. The conventional wisdom was that a business must either grow or die; it could not stand still. Harold would concede that for some businesses growth brought economies of scale. But he knew that there was a place in the economic order for businesses of all sizes. Harold was fond of playing cards, particularly poker, setback and cribbage. He applied his card sense to running the business. His strong suits were quality, reliability and a reputation for honest dealing. His weaker suits were pricing and fast turnarounds. He knew that Meriden Gravure had nothing to sell but its quality, and that the quality would be compromised if the firm or its equipment grew too large to be closely and personally supervised. The danger was that the management of an over-extended business would become bureaucratic and remote, and it would drift away from the practices that brought success in the first place.

Harold had a sense of how big an operation he could manage without sacrificing the quality of the product. Fifty employees, four 23 x 35 inch sheet-fed, single-color offset presses running one shift, and the pre-press staff and equipment to keep the presses running was about right. He steadfastly resisted the siren song of greater growth. At a panel discussion at the 1973 AAUP meetings he remarked, "From Bruce Rogers we picked up one little gem, which is the fact that supervision in a small printing office is of the greatest importance."[1] Harold then recounted the story that when Rogers was at the William Edwin Rudge plant, he was to do the press OK for the first form of every job for which he was responsible. On one occasion he was bypassed in this procedure and was very upset about it. The implication of this was that Rogers felt that the Rudge operation was getting a little too big for its own good.

Harold professed great admiration for E. F. Schumacher's 1973 book, *Small Is Beautiful: Economics as if People Mattered.*[2] Schumacher was an English econ-

Fig. 5.1. Harold in a relaxed mood. (Photo, B. A. King)

omist who was appalled by the excesses of large-scale British industrialism (the dark satanic mills). He argued for a more humane and decentralized economic order. Schumacher was preoccupied with a worldview that emphasized humanitarian, environmental and sustainability issues. Harold's view was narrower. He was concerned with the appropriate size for Meriden Gravure, but was greatly encouraged to find that not everyone thought bigger was better. He easily endorsed the proposition that people mattered.

Harold ran a tight ship. He kept a close eye on every detail of the work that was being done, in the belief that it was the details that made for the perfection of the product. Like the schoolteacher who had eyes in the back of her head, Harold could sense what was going on even when he did not have a direct view, and he knew instinctively where to anticipate trouble. On one occasion Glick was embarrassed to find that Harold was much better informed of the status of one of Glick's jobs than was Glick himself.

Harold was demanding. His subordinates accepted this because it was well understood that the company was in a demanding business. Its customers came to it with high expectations in terms of service and quality. In the eyes of its public, Meriden was only as good as its latest performance. The other aspect was that as demanding as he was of his employees, Harold was more demanding of himself. He never took time off during the business day for personal pleasures. It was his common practice to come into the shop after hours and on weekends when he could concentrate on his work in quiet and without interruptions. Harold understood that communication with and accessibility to his customers and subordinates was a part of the job, even if it interrupted his task at hand. Harold was a good, attentive listener. In discussions with a customer he would listen thoughtfully to the presentation and would hear it out to the end. Only then would he concur or put forward his own more sensible solution to the matter. With his subordinates he could be impatient, particularly with long-windedness. Harold could truly be said to have a steel-trap mind. He saw to the heart of things very quickly, and in his modesty he considered it a simple and obvious exercise of common sense that could be expected of anyone. He could become rather irritated with others not so quick-witted.

When Glick was elected to membership in the Home Club after being nominated by Harold and seconded by John, there was a dinner party at the Club for new members. Each new member was asked to get up, introduce himself and say what he did. The dinner had been preceded by a rather lengthy cocktail hour, and Glick's social discretion had become somewhat compromised. When it was his turn, he got up and ill-advisedly said, "I work for Harold Hugo. When he says 'jump' I say 'how far' " This got the intended laughs from the audience. But it really miffed Harold because he was sensitive about his reputation, and it made him appear to be a tough boss when, actually, he was a very considerate one.

Harold made himself readily accessible to employees who felt the need to speak to him. But if management needed to initiate a discussion with an employee, this was usually delegated to Les Stacey, the plant manager, or if a higher level of authority was called for, to John as General Manager. When Harold went prowling around the shop, as he frequently did, it was usually to see what work was going on and not to make small talk with the employees. Parker, after he gave up his participation in the management of the Charles Parker Co. in the mid-sixties, was regularly at 47 Billard St., but

FIG. 5.2. Meriden Gravure family picnic, American Legion pavilion, South Meriden, September 1982. (Photo, Robert Hennessey)

without many duties. He spent considerable time out in the shop and enjoyed small talking with employees.

The other opportunity for socializing with employees was at the annual company-sponsored Christmas party. This was customarily held at the South Meriden House on the Saturday night before Christmas. It involved a cocktail hour and a dinner, which gave Parker and Harold ample time to mingle. But the after-dinner activity on the dance floor or gathered around the piano to sing Christmas carols was not for them, and they invariably made an early evening exit. During John Peckham's presidency there were occasional company-sponsored picnics in addition to the Christmas party.

Harold knew how to keep his own emotions in check. He could get very exercised over what he perceived to be misconduct by some customer or supplier with whom he was working. He would dictate a letter of hot complaint to the person and put the outgoing copy of the letter in his desk drawer. A couple of weeks later he would get the letter out and tear it up. He had got his discontent off his chest without offending anyone or exacerbating a conflict.

His ability to control a crisis situation was put to the test on Monday, December 8, 1941. On the previous day, the Japanese had bombed Pearl Harbor. Meriden was in the midst of printing the second volume of *The Ledoux Collection of Japanese Prints of the Primitive Period.* This was a massive publication of classic Japanese color woodcuts from the eighteenth century. When the employees came in to work on Monday, they were understandably very upset and wanted to burn the entire job to vent their anger. Harold succeeded in calming them down by explaining that we didn't hold anything against Japanese artists who lived long ago; it was the Japanese government of the present day that Americans were justifiably opposed to. The job was not destroyed, but work on it was put aside, and only resumed and finished after the end of the war.

Harold stayed out of academic controversies. He may have had his opinions, but he was circumspect in his remarks and usually unwilling to take sides. When the Vinland Map controversy roiled Yale and was widely publicized elsewhere, Harold followed the matter closely out of personal interest and also with the hope that when Yale published the book, Meriden would get to print the illustrations, including the controversial map. The original edition came out in 1965 and the illustrations were printed at the Yale Printing Office, much to Harold's disappointment.

He also avoided political controversies. During the sixties there was much opposition to the Vietnam War in academic circles and elsewhere. Some of Harold's best friends, people like Leonard Baskin and Lyman Butterfield, were very outspoken in their opposition. Harold's great sense of loyalty extended to the nation and to officials known to him only through their media reputations. He was not inclined to vocal criticism of those who suffered sleepless nights bearing the burden and responsibility for leadership on a large scale, just as he did on a much smaller scale. He therefore avoided expressing his views to those whom he knew had strong feelings otherwise. The friendship and mutual respect between him and his dissenting friends remained undiminished.

Harold did not suffer fools gladly, but if they were important customers or potentially important customers he would suffer them reluctantly. He described his job as being paid to deal with difficult people. He said this with a certain pride because he was good at doing it. The difficult people were typically men of great ability, large ambitions and high responsibilities, but also significant insecurities. Harold was frequently in a position of trying to persuade them to do something in a different way, a way unfamiliar and uncomfortable to them. It took a great deal of tact, persuasiveness and patience to bring them to the point where they realized that the advantages of Harold's proposal outweighed its risks.

There was a saying that had currency at Meriden and elsewhere in the industry that the customer was always right even when he was wrong. Harold would always try to accommodate a customer's desires without lowering Meriden's standards. He would use all his persuasive powers to try to talk a customer out of a foolish or tasteless idea by offering a better way. Frequently he was successful, and the customer would come around to an acceptance of Harold's experience and good judgment. But there were also those who were adamant and could not be moved.

The Meriden High School Class of '27 motto, *Palma Non Sine Pulvere* (No Prize without a Struggle), was not Harold's idea and would not have suited him. He was not a self-dramatist, and he did not engage in motivational rhetoric. If he had adopted a Latin motto, it would have been something more mundane and realistic. Christopher Plantin's *Labore et Constantia* (By Labor and Constancy) might have been deemed appropriate.

Harold had many virtues. Persistence, self-confidence, resourcefulness, responsibility and integrity were important parts of his character. But the most important quality was loyalty. The company under his direction was built on loyalty: his loyalty to his employees, and their dedication and loyalty to him; his loyalty to his customers, and the trust and loyalty they returned to him; his loyalty to his suppliers, and their support and loyalty to him. Harold had a long memory for those who had stood by him in his times of need, and he was always ready to return the favor if called upon. But he also had a long memory for those who failed to make good on their commitments to him. Toward these individuals he was unforgiving. In 1942 when paper was scarce and on allocation, Harold was promised a delivery for an important job. The salesman for the paper company reneged on the agreement and sold the paper to someone else, undoubtedly for a higher price. Meriden Gravure never again did business with the salesman or the company he represented.

No bonds of loyalty were stronger than the bonds between Harold and Parker. As mentioned, Harold's decision to give up his college aspirations was in part motivated by the debt of loyalty that he felt he owed to Parker. Working for Parker, Harold was able to marry, start a family and survive the Depression in pretty good shape. He certainly merited his success, but many of his contemporaries were out of work despite their merits. It was through Parker's influence and financial help that Clarence Hugo, Harold's youngest

FIG. 5.3. Parker B. Allen, 1984. (Photo, B. A. King. Meriden Historical Society)

brother, was able to attend prep school and go on to Yale, from which he graduated in 1938.

Harold said that the only job he ever wanted was the one at Meriden Gravure. A man of his stature and accomplishment must have had many attractive job offers, but he resisted all these blandishments and was very tight lipped about what might have been offered to him. He never used these offers as a means to leverage better compensation from Parker.

Parker did not have the scholarly interests that made Harold so attractive to members of the academic community. He did not print for pleasure, nor did he collect art and books as Harold did. Nor did he choose to belong to scholarly societies. He was, however, an original member of the Columbiad Club and continued an active membership until his death.

Parker and Harold were intensely loyal to one another and were greatly dependent upon one another. If Meriden Gravure was to succeed in the scholarly publishing world, Harold would have to bring it about. Parker was acutely aware of this. Harold needed Parker's support for the direction he was taking the company and for major expenditures to upgrade or expand the facilities. Parker trusted Harold. His usual accommodating response to one of Harold's proposals was "Yes, let's work along those lines." Harold was the public face of the company, with the authority to make commitments in the company's name. This led many who did not know him well to assume that he owned the company. There were also those who assumed that of course he had earned academic degrees from one or more prestigious institutions of higher learning. This also was not the case.

Harold's relationship with Parker's wife, Elizabeth Weeks Allen, was not so comfortable. She was an active and important figure in local society. Her involvement in the affairs of the Company began when Parker went into the service in 1942, and it continued long after he returned. Elizabeth Allen was a strong-willed and controlling woman. She exercised a considerable influence over her husband. There was a creditable rumor circulating that Harold was not privy to the firm's annual financial audit because she did not want Harold to know how much money the Allen family was taking out of the business each year. In the late sixties, she became ill and her activity at the shop became curtailed. She was diagnosed with cancer and died in 1969. Presumably Harold thereafter was better informed on the finances of the company he was responsible for running.

Harold's other relationship of mutual loyalty and dependency was with John Peckham. John graduated from Princeton as a classics major in 1940 and went to work for the Princeton University Press doing production. Among those under whom he worked was P. J. Conkwright, the chief designer for the Press and one of the most respected and honored typographers working in the university press field. A mentor-protégé relationship developed. John had been at the Press about a year when Harold recruited him to come to Meriden as his assistant. This move to Meriden took John's career away from design, but he kept up a friendship with Conkwright that only ended with Conkwright's death in 1986. John hadn't been at Meriden long when America entered the war and he was called to military service. Upon his release from service he returned to Meriden and resumed working for Harold. The Korean War broke out and John, now married and starting a family, was recalled to duty. Two years later Harold got him back again, this time for good.

John was an ideal fit for Meriden. He was liberally educated; he loved books and art, and collected them throughout his life. He had a sharp eye for print quality, and a capacity for hard work and long hours. He proudly identified himself to out-of-state customers as being "from Connecticut, the land of steady habits," and his performance justified the claim. He was resourceful, responsible, energetic and completely determined that the high expectations of Meriden's customers would be fully met. He was delighted

FIG. 5.4. John Peckham, 1962. (Photo, Carl F. Zahn. Hugo Family Papers)

to be made a part of Harold's circle of book-loving friends and treasured his association with figures like Carl Rollins, Leonard Baskin, Rocky Stinehour and Joseph Blumenthal. These and many others whose acquaintance he made in the course of his work enlarged his horizons and further nurtured his interests. The enjoyment of these friendships was, as John was fond of saying, among the great non-financial benefits of working for Meriden.

John's interest in printing began in boyhood with a small hand press made by the Kelsey Press Co. of Meriden. They were sold nationwide, by mail order, complete with the necessary type, instructions and supplies for an ambitious youngster to go into business for himself printing business cards, tickets and invitations. John was only one of many who became employees or customers of the Gravure whose first experience in printing and first awareness of Meriden, Connecticut, was through the products of the Kelsey Press Co. Among the others was John's mentor, P. J. Conkwright.[3]

John, like Harold, had his own private press imprint, The Brookside Press. While Harold used the type and press equipment at the shop, John personally owned his type and press equipment and located his print shop

FIG. 5.5. Ad for the Kelsey Press Co. (Meriden City Directory, 1888. Meriden Historical Society)

in the basement of his Goodspeed Avenue home. In his retirement years John had to give up his Press. Toby Hall stepped in and acquired not only the equipment, but also the Brookside Press name, which he continues to use.

At some point in his career, John, like Harold, decided that the only job he would ever want was the one at Meriden. His ambition was to succeed Harold as president of the company. This would be a high honor but also a daunting responsibility. Harold, for his part thought John capable of filling his shoes, and would refer to John as his heir apparent. John was dependent upon Harold for a continuance of this trust and support. Harold depended on John to carry an increasing share of the responsibility for the day-to-day running of the plant. He worried that John, a hard-charging Type A personality, overworked and a bit overweight, would fall victim to a heart attack and leave a void in the upper ranks of management. This did not happen. John was up to the task and eventually succeeded to the presidency he had worked so hard to achieve.

Day-to-Day Life

There was not a lot of division of labor among the account executives working under Harold and John. Each was expected to be responsible for all aspects of a job assignment. This included estimating, purchasing, quality control and billing, as well as continuing contact with the customer. Even Harold and John did their own chores for the most part. Harold was very adept at doing figures in his head; he never kept a calculator or adding machine on his desk. What he couldn't do in his head he could do on the back of any available envelope. His results were not accurate to the penny, but were close enough to allow him to make an informed decision on the matter. In Harold's time, computers were monstrous mainframe installations run by Fortune 500 companies. Desktop computers were yet to come, but he would not have welcomed them. As stated earlier, he was not a gadget guy, and computers seemed in his mind to fall into that category. His negative attitude toward computing was influenced by his experience with a malfunctioning American Express computer that botched his credit card statement, and the aggravation he suffered in straightening it out. He ultimately became reconciled to American Express, but he never embraced the computer. Had he ever thought about it, he would have considered concepts such as artificial intelligence and virtual reality to be oxymoronic.

Unless the customer was on hand for the press run, account executives were responsible for press OKs and signed off on every sheet. The logic of this was that the account executive had the best feel for the job—what the customer's expectations were, what the potential problems might be. The critical nuances of the job, the details that made for perfection, were better executed if responsibilities were concentrated rather than diffused. There could be no passing the buck. The account executive knew his customers and in many cases had previous experience with them. He was in the best position to be the customer's advocate within the shop on matters of quality or scheduling.

Author's alterations were always a problem. They upset production schedules, threatened delivery dates and added to costs. Meriden did not like to profit from their customer's mistakes, and it tended to bill AAs at bare cost or less. If the mistake were Meriden's, as it occasionally was, it took responsibility, swallowed the cost of correction and moved on.

In cost estimating, there was always a "by guess and by golly" calculation of how smoothly the job would go and how much attention and hand-holding a customer might require. The account executive was the best judge of what this might be. The resulting job had a likelihood of smoother passage through production under this arrangement. Rapport was developed between the account executive and the customer, and the chances for a successful job and repeat business became greater.

Harold knew that the best source of new business was the repeat business of satisfied customers. The recommendation of Meriden by a satisfied customer to a colleague was a convincing and welcome endorsement.

Another practice adopted during Harold's regime and continued by John was the issuance of "savings in production," a reduction of the billing if a job went through particularly smoothly, and if the cost sheets justified it. This was a reward and encouragement for the professionalism and competency of the customer. The reduction lessened current revenue to the company, but generated long-term goodwill for Meriden. Harold always took a long view of things and was always willing to forego a small present advantage for potentially greater future advantage.

A practice that endeared customers to Meriden was the making of presentation tray cases for its customers. When a museum publication was issued, Harold would have Arno Werner make tray cases to contain the book. These were constructed of cloth over boards, with a gold stamped leather spine. They were given to the principals connected with the publication: the author or compiler of the publication, the museum director, the donor of the artwork or the supporter whose financial contribution made the exhibition and the publication possible. Such gifts were always gratefully received and highly valued.

Arno Werner, who made the tray cases, was a great friend of Harold, who became acquainted with him through the Houghton Library at Harvard, where Philip Hofer and others kept him busy with commissions to bind, rebind or encase items in their collections. Arno was born in Germany and did his apprenticeship as a bookbinder there. He came to the U.S. first in 1925, but did not permanently settle here until the late 1930s when he moved to Pittsfield, Massachusetts. He was living there when Harold first met him. Later when his wife died, he moved to Hadlyme, Connecticut, to be nearer his married daughter. This greatly facilitated his socializing with Harold, who made frequent trips to Hadlyme for lunch with Arno, often with a book in hand for which Harold needed tray cases. Harold and Arno were both dedicated to the quality and integrity of the work they were producing. With Harold it was small-scale industrial production; with Arno, it was hand-craftsmanship. Arno was very much aware of the lack of apprentice training in the crafts in the U.S. He would take on a young person for training, and a demanding and arduous apprenticeship it was. Those who survived joined the next generation of hand-binders serving the next generation of curators and collectors.

When a job was printed, it was the practice of Meriden to save the flats bearing the assembled text and illustration film, but not to save the press plates. The plates deteriorated over time, were bulky to store, and could not be changed or corrected for a later printing. The film had a longer shelf life, was less bulky and could be modified at a later date. The question of ownership of the flats was a vexing and controversial one. It was the contention of the publishers that the flats belonged to them because they had paid for the work that went into making them. Printers argued that the flats were

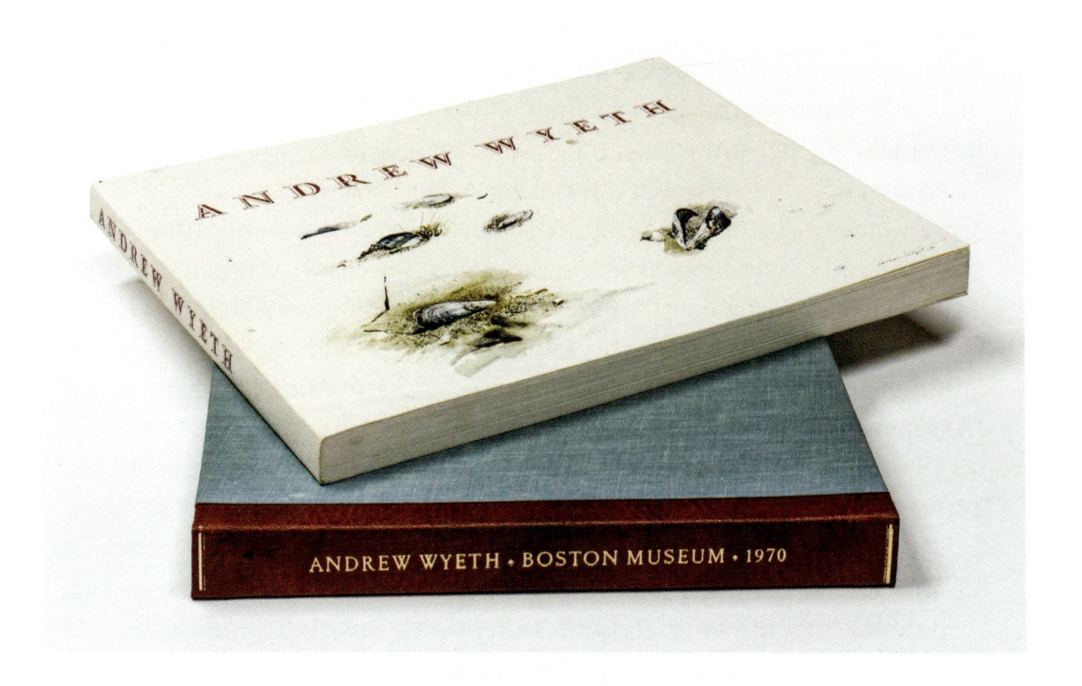

ANDREW WYETH
Cushing Road
Thomaston, Maine

Sept. 13, 1970

Dear Harold:

I have been slow indeed in thanking you for the box you so kindly sent me for the beautiful catalogue. We have it here in our living room and the blue of the box is like the sea we look on out the window.

How fine of you to have this made for me. My warmest thanks also for the fine job you did on this catalogue.

With kindest regards,

Sincerely,

Andy

Fig. 5.6. Catalogue accompanying a major loan exhibition of the work of Andrew Wyeth mounted by the Museum of Fine Arts, Boston, in 1970. Catalogue design by Carl Zahn. Tray case by Arno Werner. Fig. 5.7. Typed transcript of Wyeth's thank-you note for the tray case.

incidental to the production of the final product and remained with them. They further argued that they customarily stored the flats for the exclusive use of the publisher, and did this at no cost to the publisher. At Meriden it was argued that the flats embodied 300-line screen technology that was proprietary and nontransferable. The trade associations of the publishers and printers each had standard contract language favoring their own interpretation. For many printers involved only in very ephemeral work, it did not matter. For Meriden, printing works of lasting value with a considerable likelihood of going to reprint, it mattered a great deal. In actual practice, Meriden held out for its position where it could, but would reluctantly concede rather than antagonize a valued customer.

Many jobs were undertaken without any cost estimates because the camera work or the design and typesetting had to start, and the dimensions of the job such as the page count and the press run were as yet undetermined. A customer would commit to Meriden in the faith that Meriden would charge them fairly for the work. Meriden, for its part, would undertake the work without a contract or establishment of payment terms. It was a matter of mutual trust.

FIG. 6.1. Carl Rollins. (Photo from Columbiad Club Keepsake No. 66)

Six

A Man Among Men

Harold had a legion of friends and associates, and something of each of them rubbed off on him. A few who were particularly influential must be mentioned. Very early in his career, Harold made it his business to cultivate important people in the book world. Daniel Berkeley Updike was one, Carl Purington Rollins was another, Fred Anthoensen was a third. All three were well-established letterpress printers who would have need of the illustration-printing capabilities that Meriden offered. But equally importantly, all were committed to the practice of printing as a noble calling. Each in his own way had come under the influence of William Morris's Arts and Crafts philosophy in its American iteration. Of the three, Rollins, living in New Haven, was the most proximate to Harold. Because of this opportunity for more frequent contact, he was the most influential.

Rollins, at the time that Harold first came to know him, was in mid-career. He had earned a reputation as a thoughtful typographer, a producer of fine printing both at the Yale Printing Office and under his own imprint, and as an astute commentator on the book arts. Rollins was both Printer to Yale University and a faculty member with the rank of Professor. In this latter role, he taught a course in bibliography. In 1928 he established the Bibliographical Press to give students, primarily graduate students pursuing academic careers in the humanities, hands-on experience of typesetting and presswork.[1]

Rollins, like most of the American fine printers of the time, was not averse to machine presswork or machine-made paper. He was a traditionalist, and he admired a good letterpress impression. He was known to remark to Harold that what Meriden did was not printing but offsetting. This was said good naturedly, but he was making a point nonetheless. Ray Nash titled his 1954 tribute to Rollins *C. P. R.: Keeper of the Human Scale.* This title epitomized Rollins's attitude toward craftsmanship and the people who practiced it. It expressed his aversion to bigness for its own sake. This was influential with Harold. Harold's receptiveness to E. F. Schumacher reflected lessons taught by Rollins. Harold's doubts about numerical controls on papermaking machines, ink densitometers and other abdications of human control over mechanical processes owed much to Rollins's philosophy.

Fig. 6.2. Daniel Berkeley Updike. (Photo from *Some Aspects of Printing, Old and New*. New Haven: William Edwin Rudge, 1941)

Harold's affection for Carl extended to his wife and two daughters. After Carl died in 1960, Harold was a frequent visitor at 146 Armory Street to keep in touch with Margaret Rollins. She, for her part, welcomed his company as an old friend. He also saw a good deal of the younger of the two daughters, Caroline, who lived in New Haven and who held a number of administrative positions at Yale, some of which brought her to Meriden with printing to be done.

Harold's relationship with Updike was less active. Updike was twenty years older than Rollins and fifty years older than Harold. When Harold became acquainted with him, he was in the last decade of his life. With Updike, approval was not given gratuitously; it had to be earned by an evidence of talent and commitment. Updike initially found Hugo to be not well informed.[2] But

Harold made it his business to become well informed, and he developed a good social and professional relationship with D. B. U., the one exception being the difficulty described below. Since 1934, Harold had been making trips to the West Coast, and he served as a conduit and source of information about what was going on out there for Updike and others in the East.

In the summer of 1937, a second, updated edition of Updike's *Printing Types* was in preparation. Harold read in a trade magazine that the only changes would be sixteen pages of notes and that the pictures would be backed up. He mentioned this to Paul Standard, the New York calligrapher and bibliophile. Standard took offense at this because it made his copy of the first edition obsolete and devalued. He wrote a sharply worded complaint to Updike.[3] It was his contention that Updike and his publisher, the Harvard University Press, should issue an errata slip to be put into the first edition copies, which would make them current and preserve their value in the marketplace. Updike replied that the changes were much more complicated and extensive than that, that an errata slip would not do, and that the changes would be published only as a part of the second edition.[4] Updike wrote Harold, complaining that it was Harold's misinformation that had turned Standard loose on him.[5] Harold replied with a conciliatory letter saying that the information he had given Standard had been given in good faith and was believed to be accurate.[6] He went on to say that now that he knew the extensiveness of the changes, he entirely agreed with Updike that errata slips were out of the question and that the incorporation of the changes into the second edition was the only proper solution. Standard wrote a letter to Updike, which Updike in his letter to Hugo characterized as apologetic. Updike was somewhat mollified by this, but still clung to the notion that Harold was supporting Standard.[7] All parties wanted to bring an end to this contention, but it took a little time for the strong emotions of Updike and Standard to subside.[8] Harold wanted to be on good terms with all parties and, above all, with Updike. He felt embarrassed, and his reputation for being a knowledgeable and accurate source of information had been tarnished. In the future he would be more careful about where he got his information and how he used it.

Fred Anthoensen, at the time that Harold first came to know him in the early thirties, had established his credentials as a tasteful typographer in the conservative tradition and as a quality printer. There were a number of similarities between the two. Both had come from rather modest backgrounds; neither had a college education, yet both were more knowledgeable and erudite than many who did claim a college degree. Anthoensen's manner with his customers was similar to Harold's: a quiet demeanor and a determination to understand the customer's needs and put his knowledge at the service of his customer.

Anthoensen was twenty-eight years older than Harold, but his career ascent was slower. Walter Whitehill said of him, "Early in this century Fred Anthoensen decided what kind of printing he wanted to do. He then spent twenty-five years of unremitting work to reach the point where he could do it." [9] He joined the Southworth Press, Portland, Maine, in 1901. It was at that time a printer of religious tracts and other job printing that came its way. It was not until he became managing director in 1917 that he could begin to think of upgrading the business to the better quality bookwork to which he aspired. Ever self-critical, it was another decade before he was satisfied that he was beginning to do it.

Harold had bent the offset press intended for high-speed manufacturing to the needs and purposes of smaller-scale craftsmanship. Anthoensen had done much the same thing with the linotype machine. Built for fast and dirty newspaper production, in Anthoensen's hands it produced beautiful pages of consistent weight, without letter spacing, without excessive hyphenation, and without loose word spacing that created rivers running down the page. His success in this endeared him to Mergenthaler. Through his connections with Paul Bennett and C. H. Griffith in America and George W. Jones in Britain, he was influential in getting them to produce quality versions of such useful book faces as Caslon, Janson and Baskerville. Anthoensen thought like a French chef: if the ingredients were good, he was well on his way to a good result. His quality ingredients were his inventory of handset and machine typefaces.

In addition to design, typesetting and printing, Anthoensen had in the person of Ruth Chaplin, a superb proofreader and copy editor. He also had a close association with John Marchi, proprietor of the Marchi Bindery in Portland, Maine. He was thus able to provide complete bookmaking services except for illustration printing, which was largely but not completely supplied by Meriden. In 1934 the Press, now under Fred Anthoensen's partial ownership, became the Southworth-Anthoensen Press. In 1943 when it was entirely in his ownership, it became the Anthoensen Press.

It was through Anthoensen that Harold made the acquaintance of Walter Muir Whitehill. This was in the late thirties when Whitehill was Assistant Director of the Peabody Museum of Salem, and active in doing research and writing in maritime history. Anthoensen was printing their quarterly periodical, *The American Neptune*, as well as their book-length publications. After distinguished naval service in Washington during World War II, White-

FIG. 6.3. Fred Anthoensen. Undated photo reproduced in collotype by Meriden for *Fred Anthoensen: A Lecture by Walter Muir Whitehill given at the Composing Room, New York City, 23 February 1966.*

hill returned to Boston as Director and Librarian of the Boston Athenaeum. But this was far from his only occupation. He held a faculty appointment at Harvard, which included classroom teaching, administrative and advisory duties. He was active in local antiquarian and preservationist groups and with the Bostonian Society, whose collections and exhibitions were devoted to the history of the city. He was an indefatigable lecturer, reviewer, writer, committeeman, trustee, promoter of worthy civic causes and arbiter of cultural life in Boston.

As an undergraduate at Harvard, Whitehill had taken George Parker Winship's Fine Arts 5e, "History of the Printed Book." He knew good printing, and he cared greatly about it. He was for a time the editor of *Daedalus*, the scholarly publication of the American Academy of Arts and Sciences, and had occasion to use Meriden when its publications in art history were featured. He was later editor of publications at the Colonial Society of Massachusetts, whose publications were printed by Anthoensen in collaboration with Meriden, and subsequently by Stinehour in collaboration with Meriden. He wrote the two-volume *Museum of Fine Arts, Boston: A Centennial History*, published in 1970 and printed by Meriden. He was the author of *Boston, a Topographical History* (1959), published by the Harvard University Press. This chronicled the changing face of the city from its founding in 1630 down to the present day. It was printed with a generous measure of illustrations by Meriden. As Boston was undergoing rapid change in the postwar decades, updated editions were printed in 1963 and 1968. It was Whitehill

FIG. 6.4. Walter Muir Whitehill at Dyke Mill, Montagu, Mass., at Carl Rollins's 1954 birthday party. (Photo, Nancy Price. Hugo Family Papers)

who recommended Anthoensen and Meriden to Armistead Peter III for the printing of his book, *Tudor Place,* whose production is among the adventures in printing described by Peckham.

Ray Nash was an important figure in the graphic arts scene, not so much as a printer, but as an educator and writer on the graphic arts. In the thirties he taught graphic arts courses at the New School for Social Research in New York before leaving the city in 1937 to take up residence in Vermont and to teach at Dartmouth. His philosophy of teaching was developed at the New School and was carried on to Dartmouth, but it much resembled what Rollins was doing at Yale in his bibliography course and with the Bibliographical Press. Both were original members of the Editorial Board of *Print Magazine,* founded by Edwin Rudge III in 1940. Like Rollins, Nash emphasized a "hands-on" approach to teaching. He required his students to practice close observation and careful study of the works of graphic art themselves, rather than study photos, lantern slides or printed reproductions of the material. Nash's tribute to Rollins was his 1954 exhibition at Dartmouth of Rollins's work and the accompanying publication, the aforementioned *C. P. R.: Keeper of the Human Scale.*

Nash's insistence that study must be from the originals was very similar in spirit to Harold's determination that reproductive photography should be done directly from original sources. Each came to his conclusion independently of the other. But they concurred that authenticity matters; this became a part of the bond between them. Beyond this, they had a common circle of friends and associates, and they held many memberships in common.

Although Harold and Nash had much in common, they also had differences in lifestyle. Nash had grown up in rural Oregon and was devoted to country living. He actively farmed his Vermont property. During World War II, Harold as a patriotic obligation kept a victory garden at 423 Westfield Road in Meriden. But the minute the war was over, he dropped his shovel and hoe and gave up tilling the soil. Thereafter the garden became the sole responsibility of Marge Hugo, who was the avid horticulturalist in the family.

Nash presented printing and graphic art to his students as a significant part of the fabric of civilization, appropriate to study in a liberal arts curriculum. It was not intended to be professional training in technology or aesthetics. But some of his students were so taken with the experience that they made a career in teaching the book arts, publishing, graphic design or curatorship. These figures are celebrated as the legacy of Ray Nash. Among them was Roderick Stinehour. Unlike many of Nash's students, Stinehour was not a blank slate with respect to printing when he came to Dartmouth. His first exposure to printing

Fig. 6.5. Ray Nash and Harold Hugo at Dyke Mill, Montagu, Mass. at Carl Rollins's 1954 birthday party. (Photo, Nancy Price. Hugo Family Papers)

had come at the Bisbee Press in Lunenburg, Vermont. It was Mr. Bisbee's suggestion that he seek out Professor Nash at Dartmouth.

Rocky was interviewed by Nash, and on Nash's recommendation, he was admitted to Dartmouth in the fall of 1948. He eagerly took every course offered by Nash and became a Nash protégé. Nash gave him the opportunity as a senior honors project to curate and organize an exhibition. Rocky's chosen subject for the exhibition was the work of Ned Thompson's Hawthorn House in Windham, Connecticut. Thompson was a country printer, operating on a modest scale, producing small tasteful jobs for discriminating clients, and for himself as publisher. This was something of interest to Rocky; it was something to which he himself could reasonably aspire. Thompson's signature typeface was Bulmer. It is not coincidental that Rocky adopted Bulmer as his first workhorse typeface and used it for *PaGA* and many other early publications of the Press.

Nash put Rocky in touch with Harold, who had a good collection of Hawthorn House material. Harold, for his part, was entirely willing to lend his collection to someone who came well recommended by Nash. This was the small beginning of a long and active friendship and fruitful collaboration. Had they not met through Nash at Dartmouth, they certainly would have met soon thereafter. Rocky returned to Lunenburg after his 1950 graduation, and established himself as a typographer and printer with a credible claim to "print books better than is ordinarily done." It was this activity that sustained the relationship throughout Harold's lifetime.

Stinehour, fifteen years younger than Harold, was nonetheless of another generation. Harold had entered into the adult world at the beginning of the Great Depression, and learned its lessons of caution and wariness. Rocky entered the adult world by way of wartime military service, with all the risks and hazards that that entailed. He was a member of what has come to be called "the greatest generation," now home safely and victorious, but much sobered and matured by the experience. He was optimistic, entrepreneurial and self-confident in spirit. His first civilian enterprise, the flying service, failed, but the printing venture prospered. It soon grew well beyond the "one man with assistance from his wife and a single employee" model of Ned Thompson.[10]

As the Stinehour Press added payroll and grew in its capabilities, it was able to take on larger projects, many of which required illustrations from Meriden. Several of these early jobs were done in collotype. But collotype was on its way out, and most subsequent jobs were done in offset. Harold recognized Rocky as the most energetic and promising of the younger generation of craftsman printers. Soon Rocky was being introduced into the Nash/Hugo circle of friends and associates, and he was being invited to membership in many of the organizations and clubs to which they belonged.

Prominent among those in the Nash/Hugo circle was Thomas R. Adams, of the John Carter Brown Library in Providence, Rhode Island. Tom was the son of Randolph Adams, the Librarian of the Clements Library at Ann Arbor, Michigan to whom Harold had introduced himself by way of his publication of *Early Connecticut Printing.* In 1957 Tom Adams had succeeded Lawrence Wroth as Librarian of the JCB, the research library dedicated to the collecting and study of the age of exploration and colonization of the Americas. Harold had caught Wroth's attention in 1938 with the publication of *Loyalist Operations at New Haven* (see Fig. 2.10), which Wroth mentioned favorably in his column, "Notes for Bibliophiles," in the *New York Herald Tribune.*[11]

Harold's business connection with the JCB began in 1941 when Wroth published a facsimile of the rare 1673 Augustine Herrman map, *Virginia and Maryland.*[12] This was done at full size in collotype at Meriden and was so successful that it went through two subsequent printings. During Adams's

Fig. 6.6. The guest list at the wedding of Rocky Stinehour's daughter Ann in 1974 included many of Rocky's professional friends. Gathered here are Harold Hugo, Freeman Keith, David Godine, Rocky Stinehour, Ray Nash and Joseph Blumenthal. (Photo, Stinehour Editions)

tenure Meriden produced a number of important publications for the JCB. Of these the most monumental (to use one of Peckham's favorite words) was *The Blathwayt Atlas.*

The Blathwayt Atlas was a collection of forty-eight maps, dating from the late seventeenth century, reflecting British exploration and overseas trade interests. It has been called the first atlas of the British Empire. The maps, some printed and some in manuscript, were each reproduced at their original image size. Several were so large that they had to be printed in two sheets and be pasted together. One was printed in collotype and hand-colored. Thirty-nine of the maps were done in collotype uncolored. This was among the last of the great collotype jobs to come off Meriden's presses. The other eight maps were printed in offset in four-color process. In his Foreword to the map volume, Adams writes, "the quality of the reproductions is the result of the care and attention of Mr. Harold Hugo, president of The Meriden Gravure Company.

Fig. 6.7. Tom Adams (right) with John Maggs, London, 1967. (Photo provided by Mrs. Thomas R. Adams)

He personally carried almost all of the original maps, one by one, to and from Meriden, where all the photography was done."[13] The publication, advertised by prospectus in 1962, was finally issued in 1970 when Jeanette Black's extensively researched text volume was completed.

Harold was a founding member of the Associates of the John Carter Brown Library and was a regular attendee at the annual meeting of the associates, along with Jim Barnett who was a Brown University graduate. They would stay at the Wayland Manor (often referred to as the Wayward Manor), and this would be the gathering place for a nice dinner before the business meeting, and maybe a nightcap afterwards, hosted by Harold and Jim and attended by various other Associates. Tom and Harold shared accommodations on several of the Grolier Club's bibliographic pilgrimages to Europe.

The other cherished friend in Rhode Island was John Howard Benson, the master stone carver from Newport. He was the proprietor of the John Stevens Shop, founded in 1705 and reputed to be the oldest continuously operating business in the United States. Richard Benson has explained the circumstances regarding their first collaboration. "My father first met Harold when the Gravure reproduced the eighteenth century account book for the John Stevens Shop. Harold had this done in collotype, and my father loved the job and became very close to Harold afterwards." [14] The second noteworthy project, the Arrighi writing book, was issued in 1954, two years before Benson's death. This is one of the books taken up by Peckham in *Adventures in Printing*. The younger

FIG. 6.8. John Howard Benson and Carl Rollins at Dyke Mill, Montagu, Mass. at Rollins's 1954 birthday party. (Photo, Nancy Price. Hugo Family Papers)

Benson (who was 12 at the time of his father's death) further remarks, "When my father died Harold was the first person to show up at our house—he heard the news of his death, got in his car and drove to Newport—I remember how deeply moved we all were by his presence at our house."

Harold's connection to the New Haven community began with Carl Rollins and his family, but as time went on extended far beyond them. It was almost exclusively individuals who were connected with Yale, either through the Yale University Press and Printing Office, the Yale faculty, or the Yale administration, particularly that of the Library. Yale was a fertile ground for the recruitment of members of the Columbiad Club. James Boyden, Production Manager, Yale University Press, Greer Allen, University Printer, and later his wife, Sue Allen, a research scholar in the book arts, were members, as was Roland Hoover, Greer Allen's successor as University Printer. Typographer John McCrillis was a long-time member, and Alvin Eisenman, under whom McCrillis worked, was a member briefly. Howard Gralla, another active member of the Columbiad Club, taught at Yale in addition to designing and producing many publications carrying a Yale imprint. It was through Gralla that Meriden did work for Edward Tufte, a Yale faculty member whose teaching and published research integrated the disciplines of economics, statistics and graphic design. Unlike most academics, Tufte insisted on self-publishing his research at his own expense (and ultimately to his own profit) rather than submitting it to a university press for publication. Under his imprint, The Graphics Press, Tufte's *The Visual Display of Quantitative Information* went through ten substantial printings in the decade after its 1983 debut. These were done at Meriden prior to the consolidation of 1989, and subsequently in Lunenburg.

Through his memberships in the Associates of the Yale Library and the Grolier Club, Harold had contact with James Babb, Herman Liebert, Kenneth Nesheim, Gay Walker, Stephen Parks and Marjorie Wynne. Archie Hanna was a frequent companion and often accompanied Harold to Boston for the monthly meetings of the Club of Odd Volumes. During Hanna's curatorship of the Coe Collection of Western Americana, Meriden printed many of its annual keepsakes.

Harold followed with interest the career of William S. Reese, founder and guiding spirit of the antiquarian book dealership William S. Reese Co., located in New Haven. Their common institutional enthusiasms were the Yale Library Associates and the American Antiquarian Society. Chester Kerr, Director of the Yale University Press, was an important figure in the Yale community and in the world of scholarly publishing. Harold cultivated his connections with Kerr and with his colleagues in the editorial, design and production departments at Yale University Press.

Harold enjoyed a long, cordial friendship with Yale graduate, trustee and benefactor Wilmarth S. Lewis. Lewis devoted his life to the collection, study and publication of the work of Horace Walpole. This inevitably expanded to include not only Walpole himself, but the whole panorama of Georgian England. Harold was on many occasions a luncheon guest of Lewis's, not usually in New Haven, but at Lewis's gracious home and extensive library in Farmington, Connecticut. Meriden printed illustration inserts for many of Lewis's books, all dealing with Walpole or with Lewis's pursuit of him.

Fig. 6.9. Wilmarth Sheldon Lewis in his library, 1966. (Photo attributed to W. F. Miller & Co. and provided by The Lewis Walpole Library, Yale University)

Harold was particularly proud of his part in facilitating the publication of *The Beggar's Opera by Hogarth and Blake*, a collaboration between Lewis and Philip Hofer that is detailed in *Adventures in Printing*.

The Wesleyan academic community, though smaller than Yale, was no less vibrant, and Harold had many connections there. He was a member and benefactor of the Friends of the Wesleyan Library and the Friends of the Davison Art Center. A number of faculty and high-ranking staff were members of the Columbiad Club: Willard Lockwood, Director of the Wesleyan University Press, and Raymond Grimaila, his design and production chief; Russell "Butch" Limbach, the much-loved teacher of the studio art courses; and Wyman Parker, the University Librarian. Much of the printing that Meriden did for Wesleyan was for the Wesleyan University Press, which was particularly active in publishing contemporary American poetry, American decorative arts, the theater, and regional and local history. In some situations the illustrations were printed to insert. In other more lavishly illustrated publications, Meriden would do the entire book, with illustrations integrated into the text throughout. Meriden printed a photographic book and a number of catalogues for Wesleyan grad Philip Trager and several publications celebrating Wesleyan's 1981 Sesquicentennial. Meriden occasionally did announcements and exhibition catalogues for the Davison Art Center.

Harold also had friends on the faculty whose company he particularly enjoyed. One was Paul Horgan, a Pulitzer Prize-winning author of books on the Southwest who held a faculty appointment as writer-in-residence. Another was Joseph Reed, professor of English with wide-ranging interests in literature, film, the visual arts and contemporary culture. Horgan and Reed were frequently Harold's guests at the Home Club. Hospitality was reciprocated, and extended to Peckham and Glick as well. Reed's creativity, imagination and originality found expression in various art forms: drawing, etching and small-scale sculpture. An invitation to lunch at Reed's Lawn Avenue home, hosted by Joe and his novelist wife, Kit Reed, was often the excuse for the unveiling of Joe's latest artistic inspiration. Other staff and faculty members who were in frequent contact with Meriden were Sam Green and David Schorr of the Art Department, Arthur Wensinger (Jerry to his friends), Ellen D'Oench, curator of the Davison Art Center for many years, and William Van Saun in the Publications Office.

Leonard Baskin created powerful images and was himself a powerful personality. Harold met him through Clarence Kennedy, Baskin's colleague on the Art Department faculty at Smith College. In addition to his work in sculpture, prints and drawings, Baskin operated the Gehenna Press, and his interest in printing and typography was one of the connections between them. Another bond they shared was an attitude of unflinching integrity toward his work. Harold was a frequent visitor to Baskin's home at Fort Hill in Northampton, Massachusetts and at his studio on Little Deer Isle, Maine, where he worked summers.

Meriden did illustrations for several Gehenna Press publications. *Four Portrait Busts By Francesco Laurana: Photographs by Clarence Kennedy with an Introductory Biographical Essay by Ruth Wedgwood Kennedy* appeared in 1962. *Rembrandt's Book of Tobit* by Julius Held followed a year later. In 1969 Meriden printed Baskin's drawings for Thomas Bergin's translation of *The Divine Comedy*, published by Grossman in three large quarto volumes. The text was set in Giovanni Mardersteig's Dante type, and was designed and printed in Verona, Italy by Martino Mardersteig at the Stamperia Valdonega. Meriden also reproduced Baskin's forty-nine drawings for an illustrated edition of *The Iliad* in the Richard Lattimore translation published by the University of Chicago Press in 1962.

Harold was the subject of a portrait by Baskin, a relief bust in bronze, done in the summer of 1970 when he visited Baskin at his Little Deer Isle studio. Of the two known copies, one is at the Davison Rare Book Room in the Olin Library, Wesleyan University. The other copy was Harold's copy, which was purchased after his death by Veatchs Arts of the Book. In 2008 a subscription, spearheaded by Toby Hall, was taken up among Harold's friends to acquire the portrait for the Boston Athenaeum, where it is now on permanent display.

Baskin championed the work of his contemporary, Rico Lebrun. A great

Fig. 6.10. Leonard Baskin at Little Deer Isle, Maine, 1974. (Photo, B. A. King)

FIG. 6.11. Parker and Harold at Baskin's summer home and studio, Little Deer Isle, Maine, 1974. (Photo, B. A. King)

draftsman and very much a kindred spirit, Lebrun was less well known and less promoted in the art market. Through Baskin's influence, Meriden printed *Drawings for Dante's Inferno* by Rico Lebrun, published by the Kanthos Press in 1963. This was not the entire text, but excerpts from it in the translation of John Ciardi that inspired the illustrations. Ciardi wrote a Foreword, and Baskin provided a Note on the Drawings. The format was folio size, which allowed Baskin to design in his monumental style. The Stinehour Press set it in their largest sizes of Bembo.

This work was almost finished and the extra illustrations that were to be laid into the book were being trimmed to final size on a quiet Friday afternoon in November. But work was disrupted with the report that President

Fig. 6.12. The Wesleyan copy of the Baskin Relief Portrait of Harold Hugo, 1970. It measures about 24 inches high. (Photo, Wesleyan University Library, Special Collections & Archives)

Kennedy had been shot in Dallas. When the news circulated, the operator of the paper cutter was so upset that he miscut and spoiled some of the sheets. Most of the employees left work early and went home or to church, each to be in his or her own grief.

Baskin was also the art editor of the *Massachusetts Review*, a literary quarterly published by the University of Massachusetts, Amherst. This gave him an opportunity to promote the art of little-known artists who were, in Baskin's mind, deserving of greater recognition. Meriden regularly printed these illustrated sections for *MR*.

Harold's overseas friends were largely British. American printing, publishing, typography and bibliographic studies had from its beginnings been strongly influenced by British practices. Harold reflected the anglophilism of many of his mentors, friends and colleagues. Harold had no facility in foreign languages, and he was particularly uncomfortable with francophones. The British Isles were "just across the pond"; the continent of Europe was an ocean away. He was well acquainted with many of the leading figures in British graphic design, book production, publishing and bibliographic scholarship including Will Carter, John Dreyfus, Rowley Atterbury, Sir Frank Francis, Nicolas Barker, Sebastian Carter and Vivian Ridler.

Ruari McLean was Harold's closest friend in Great Britain. The hearty, good-natured Scot was a leading book designer in the postwar period, the author of a number of authoritative works on the history of book design and editor of the influential design journal *Motif* during its nine-year existence. His article "The Reproduction of Prints" appeared in the March 1987 issue of the London periodical *Print Quarterly.*[15] Writing to a British readership, it surveyed the various processes in use in the commercial reproduction of monochromatic images over the past hundred years. These included line cuts and photoengravings printed in relief, intaglio processes such as gravure, collotype and offset lithography. It also included mention of new processes such as photopolymer plates, ink jet printing and electrostatic (i.e., Xerox) printing. Relief printing, intaglio and collotype were dismissed as being either inherently inferior in quality or prohibitively expensive or both. The newer technologies were promising but unproven. This left offset lithography as the process of choice. He writes, "Many of the improvements which have made photo-litho-offset the most suitable process for the commercial reproduction of prints are due to one firm in the United States—the Meriden Gravure Company . . . and indeed to one man . . . Harold Hugo." He continues, "They were gradually recognized as the best printers in this field not only in America but in the world." The conclusion of his article exhorts his countrymen to consider the following: "It is said that French cooking is as good as it is, because Frenchmen complain loudly if it isn't. There may be a lesson here for our printers, and for those who use their books." In this and other publications McLean preached the gospel of Hugo: work from originals and master 300-line screen offset.

Harold was elected an honorary member of the Double Crown Club of London in 1967. The common interest of the Club was bibliophilic, and its members included all of the most important figures in the world of books in Britain plus many honorary members from abroad. Harold was invited to address a dinner meeting in 1971. He spoke on "Graphic Reproductions at

FIG. 6.13. Ruari McLean. (Photo provided by David McLean)

The Meriden Gravure Co." Meriden was frequently asked to produce exhibition catalogues of loan exhibitions from England. Through his connections to important figures in the museum and library world, he was sometimes able to arrange to photograph directly from the originals. Where that was not possible, he would establish stringent quality standards for photographs to be supplied to him.

Memberships

Harold, as Rocky wrote in his Introductory Note to the printed edition of *Adventures in Printing*, was "the most clubbable of men." In addition to the Columbiad Club, which he helped to found, Harold was invited in 1950 to membership in the Grolier Club, which stimulated his interest in collecting its publications. He also became a member of the Century Association in New York and several Boston organizations: the Society of Printers and the Club of Odd Volumes, the St. Botolph Club and the Massachusetts Historical Society. He was a member of the American Antiquarian Society, Worcester; the Yale Library Associates, and the Associates of the John Carter

Brown Library in Providence. He was a Fellow of the American Academy of Arts and Sciences and a Life Fellow of the Metropolitan Museum of Art. His membership in the Bibliographical Society of America dated from the early thirties, and his membership in the Bibliographical Society (U.K.) dated from the late thirties. The members of these scholarly and bibliophilic organizations were important figures in the publishing and academic world who were among the company's most valued customers. Of all these organizations, the two closest to his heart were the Columbiad Club and the Club of Odd Volumes. He was a regular attendee at the monthly dinner meetings of both. The COV also had Saturday luncheons of a more informal nature that Harold particularly enjoyed. Two hundred and fifty miles round trip was, he admitted, a long way to go for lunch, but good food and drink and the stimulating conversation with the others present at table made it all worthwhile. The Saturday lunch was usually preceded by a drop-in and browse at Goodspeed's Book Shop, after which George Goodspeed would join him in a walk down Beacon Hill to a well-stocked bar and Mme. Robineau's well-served table at 77 Mt. Vernon Street.

Locally, Harold was a long-time member of the Home Club of Meriden. During his term of office as president he put through an expansion of fa-

FIG. 6.14. Photo of Harold, taken in 1960 when he was elected President of the Home Club. It hung in the Men's Bar of the Club with photographs of his predecessors and successors in that office. (Hugo Family Papers)

cilities that was derided by some shortsighted members as "Hugo's Folly," but which proved to be a significant and valuable improvement. He was an honorary director of the Meriden-Wallingford Manufacturers Association, a corporator of the Meriden-Wallingford Hospital and served on the advisory board of the Connecticut Bank & Trust Co. Harold also served as a Trustee of Old Sturbridge Village, where he gave particular attention to strengthening its finances and to building the collections of the research library. The Research Library Society established a book prize for the best published contribution to the understanding of rural New England culture in the early republic. It is named in memory of Harold.[16] He also served as a trustee of the Klingberg Family Center, an orphanage turned social services organization in New Britain, Connecticut, founded by a Swedish-American pastor, the Reverend John Eric Klingberg.

Harold never joined one of the many service clubs in Meriden. They were demanding of his time, and the exuberant bonhomie of the Rotary or the Lions Club was a bit beneath his dignity. He never served in any municipal capacity, either as an elected or appointed official. While the city might have benefitted from his good counsel, Harold absolutely disdained politics and politicians whom he knew first hand. They were, in his opinion, unreliable, pandering opportunists, and he wanted no part of them. One political figure he did respect, however, was Norman Thomas, six-time unsuccessful candidate for President on the Socialist ticket. He recalled several occasions seeing Thomas on the street in Princeton. It is doubtful that he ever voted for Thomas. He would have voted Republican out of his own conviction and out of support for Parker, who was conservative in his views. But, ideology aside, he admired Thomas for his honesty, determination and steadfast commitment to principles.

The Congresses

In addition to the many formal organizations, there were the congresses. The congresses were occasional weekend retreats of a distinctly secular and non-spiritual character. Julian Boyd hosted the Bear Lake Congress (more properly known as the Bear Lake Historical, Typographical, Gastronomical, Marching, Singing, and Grouse Poaching Congress) at his vacation home in the Poconos. Lyman Butterfield convened the Glades Congress at his summer place in Minot, Massachusetts. Walter Whitehill's Camel's Hump (or Camel's Rump) Congress met in the Green Mountains of Vermont. Herb

Farrier summoned Congressmen to the Biddeford Pool Congress in Biddeford Pool, Maine. The Biddeford Pool Congress met regularly in September (hurricane season in New England, but no matter) and sustained itself longer than its sister organizations. Harold was a member of all of these congresses, as were Parker, Jim Barnett and, later on, John Peckham. Each convener included a number of his most congenial friends and associates as delegates to his congress. Fred Anthoensen and John Marchi of the Marchi Bindery in Portland were members of the Biddeford Pool Congress, but declined to participate in the congresses outside of their home state. P. J. Conkwright, Chiang Yee and Francis L. Berkeley, all of them in the academic world, were members of the Bear Lake Congress.

Some congresses had their own handsome congressional letterheads. Each congress had its officers, most of whom held ceremonial titles without any duties. The exception to this was the Quartermaster, whose weighty responsibility was to provide food and drink for the assembly. The favored libation was Fairfax County whiskey and the preferred food that of the locality. At Biddeford it was lobster; at Minot, shellfish. The order of business did not vary much from one congress to another: an abundance of food and drink, with stories and spirited conversation into the wee hours. It was not unusual for the Meriden delegation to return home a little bit the worse for wear following a Biddeford weekend.

Frank Wardlaw, Director of the University of Texas Press, was a guest Congressman at one of these venues. He had on many occasions joined other Texas luminaries such as Walter Prescott Webb and Roy Bedichek at J. Frank Dobie's ranch at Paisano. These gatherings were conducted in a spirit similar to the Northeastern Congresses, except that in the matter of spirits, the Texans preferred Jack Daniels to Fairfax County. Thus it was that Paisano was accorded congressional status. Harold was a Paisano delegate on at least one occasion, as was Walter Whitehill.

The Indoor-Outdoor Athletic Club was an informal group of Meriden men mostly known to one another from the Home Club. It met on Sunday mornings (while the wives and children were going to church) for a brisk walk, after which the indoor activity commenced in Parker's basement. It consisted of talk, storytelling and a suitable liquid refreshment to promote the discourse. Each member of the Indoor-Outdoor AC had an animal name beginning with the same letter as the member's family name. Parker Allen, the acknowledged leader of the group, was the Autocratic Ape. John Peckham, an active member and enthusiastic athlete, was the Panda. Petty fines were imposed for the failure to address a member by his animal name.

Bear Lake

HISTORICAL, TYPOGRAPHICAL, GASTRONOMICAL,
MARCHING, SINGING,
AND GROUSE POACHING CONGRESS, INC.

OFFICERS

Julian Parks Boyd, *Chairman*, Princeton, New Jersey
Lyman H. Butterfield, *Treasurer*, Boston, Massachusetts
Walter Muir Whitehill, *Clerk*, Boston, Massachusetts
Pleasant Jefferson Conkwright, *Typographer*, Princeton, New Jersey
Everett Harold Hugo, *Quartermaster*, Meriden, Connecticut

MEMBERS

The above Officers, as in the *Mexican Army*

HONORARY MEMBERS

Gilbert Stuart McClintock, Wilkes-Barre, Pennsylvania
James P. Harris, Wilkes-Barre, Pennsylvania

GUEST

Henry Kretchmer, Gouldsboro, Pennsylvania

HEADQUARTERS: Bear Lake Association, Gouldsboro, Pennsylvania
EMBLEM: Three Grouse Feathers* CHAMPAGNE: G. H. Mumm & Co.
SONG: Hail Columbia BEEF: Black Angus
FLOWER: Rhodedendron WHISKEY: Fairfax County
SCRIPTURE: H. L. Mencken: *Christmas Story*

*Because of poor Poaching conditions in the Fall of 1955, the Congressional Emblem is represented with only one specimen.

Fig. 6.15. Membership roster of the Bear Lake Congress. (Hugo Family Papers)

The group met regularly from the first Sunday after the Yale-Harvard game through the winter to Palm Sunday. Harold was a member but participated only occasionally, and his animal name has been lost to posterity. He was not much interested in physical exercise, even if the exertion might be no more than a brief walk. His chief recreation was a sedentary but competitive game of cribbage with his friends at the Home Club.

Harold for all his underlying seriousness was good natured and companionable. He could take a joke at his own expense in stride and not be offended; he could easily be persuaded to join in a practical joke at the expense of a friend. The most notable example of this was the production of the Binney & Ronaldson type specimen sheet. In the 1950s Harold was returning from a bibliophilic gathering with P. J. Conkwright and the book dealer

Fig. 6.16. The Biddeford Pool Congress in session, 1957. Left to right: Allen, Anthoensen, Marchi, Whitehill, Hugo and Barnett. Note that none are empty handed. Not pictured is Herb Farrier, who probably took the picture. (Illustration from *L'Affaire Farrier,* The Typophiles, 1963)

Dick Wormser. Conkwright was studying early American type foundries and believed that there might be a Binney & Ronaldson type specimen sheet earlier than the one known to have been issued in 1809. After Conkwright had been dropped off at Princeton, the others continued on to Connecticut. Wormser, a mischievous practical joker, proposed that they concoct a bogus broadside. Harold readily joined the conspiracy and agreed to use Meriden's resources to accomplish it. Twenty copies of a Binney & Ronaldson type specimen sheet dated 1802 were printed on an odd lot of old ledger paper and included specimens of Persian and Tibetan typefaces. The title was printed in Underwood Elite, and the date was printed with a numbering machine. Everyone enjoyed the joke, Conkwright included. The demand for the broadside among Harold's friends and associates could only be met by a second printing of another fifteen copies.

Family Life

Harold married in 1935. Marjorie Ekberg and Harold Hugo were a logical match initially; they shared the same ethnic background and Swedish Baptist religious affiliation. Their daughter, Nancy Ellen Hugo, was born the following year. Their son, Gregg Hugo, nicknamed Tuck, was born nine years later. Because their children were nine years apart, parental duties extended over a considerable period of time. These largely fell to Marge, given Harold's long hours at the shop and many absences on business travel. Harold was extroverted and intellectually curious. Marge's interests were her family, her household and her garden. Harold had a greenhouse addition to the house built so she could garden year round. She was not a very social person and was not greatly engaged in activities outside the home. She was an unenthusiastic participant in the several family vacations in Maine. This was in part because it took her from home and garden and because travel time to Maine was lengthened by stops at the Anthoensen Press, and perhaps other business calls. After the children were grown, Harold's vacation travel was no longer a family affair, and Marge was no doubt happy to be left at home to pursue her own pleasures. As noted previously, Harold did a Grolier Club junket with Tom Adams. His 1967 trip to classical sites in Greece and Turkey was in the company of Lyman and Betty Butterfield. His visits to Leonard Baskin at Little Deer Isle in the seventies were with Parker, and a trans-Canada train trip was done with his bachelor brother, Clarence. The fishing expedition to Richardson Lake in Maine in the early

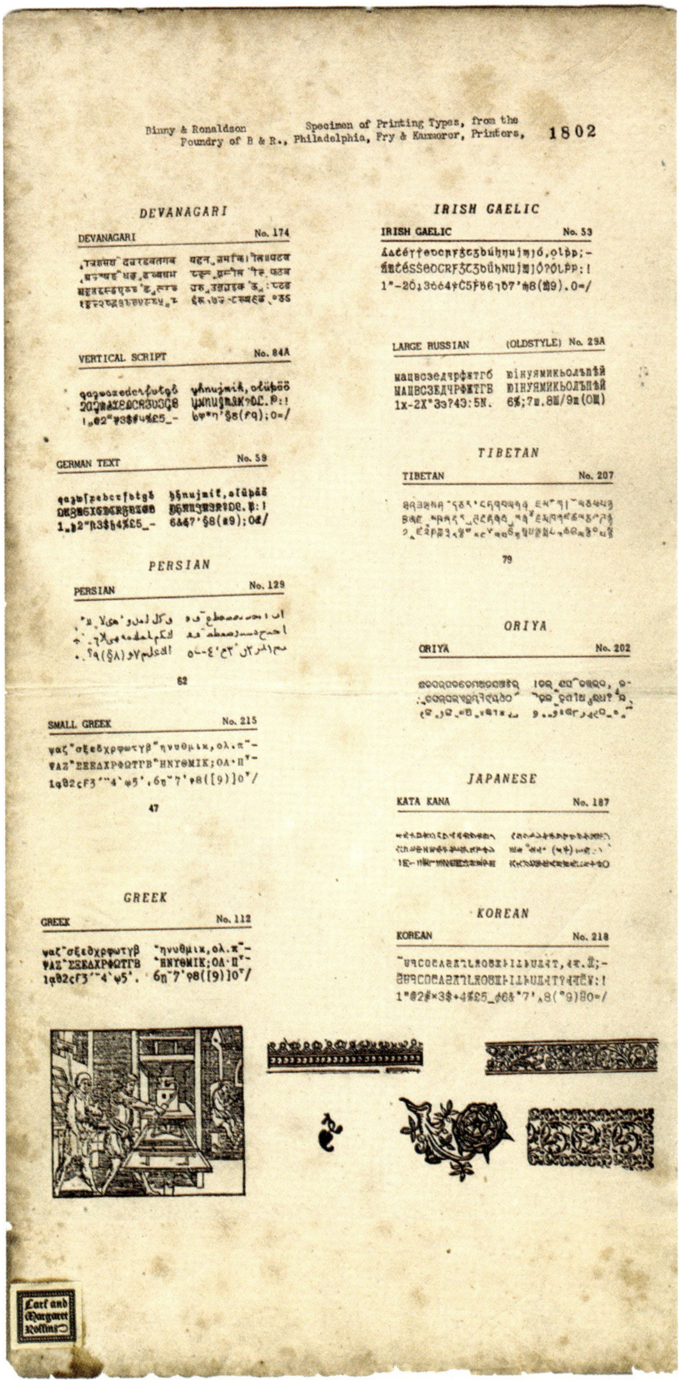

Binny & Ronaldson Specimen of Printing Types, from the Foundry of B & R., Philadelphia, Fry & Kammerer, Printers, 1802

DEVANAGARI

DEVANAGARI No. 174

IRISH GAELIC

IRISH GAELIC No. 53

VERTICAL SCRIPT No. 84A

LARGE RUSSIAN (OLDSTYLE) No. 29A

GERMAN TEXT No. 59

TIBETAN

TIBETAN No. 207

79

PERSIAN

PERSIAN No. 129

ORIYA

ORIYA No. 202

62

SMALL GREEK No. 215

47

JAPANESE

KATA KANA No. 187

GREEK

GREEK No. 112

KOREAN

KOREAN No. 218

Carl and Margaret Rollins

Fig. 6.17. Carl and Margaret Rollins's copy of the bogus Binney & Ronaldson 1802 Type Specimen Sheet. (Photo courtesy of Carl Purington Rollins Papers, Robert B. Haas Family Arts Library, Yale University)

FIG. 6.18. Gregg Hugo in his office, 1978. Note the 1906 rendering of the plant (Fig. 1.6) hanging on the wall at the left. (Photo, B. A. King. Meriden Historical Society)

sixties, which was chronicled by Ruari McLean,[17] was done with Jim Barnett and Harold's son, Gregg.

Harold did most of his business entertaining at the Home Club, but occasionally entertained at home if a guest had to be put up for the night. Rocky Stinehour was a frequent guest at 423 Westfield Road after the merger. In these instances Marge was a dutiful hostess, but not an involved participant in the social activity. But when Chien-Fei and Chao-Wen Chiang arrived as immigrants in a strange land with a babe in arms, it was Marge who was so helpful and supportive of them and who devoted so much of her time to familiarizing them with domestic life as practiced in this country.

Harold was proud of his children. Nancy was a top student, went to Radcliffe and graduated in the class of 1958. She spent several of her high school and college summers at the shop, working in the retouching department. After serving for a year as a researcher at the Massachusetts Historical Society, she embarked on a long and distinguished career teaching high school English in New Canaan, Connecticut. Nancy was an avid photographer and adventurous traveler, and made good use of her free summers in these pursuits. Her brother, Gregg, was not strong academically, but he inherited his father's work ethic, common sense and organizational skills. He also worked high school summers at the shop. After graduation and military service, he was happy to return to his hometown and join Meriden Gravure full time. Harold was pleased that his son thought that the Gravure was a good place to work, and was glad to have him. Gregg worked his way up in increasingly responsible positions in production and, at the time of the consolidation, was plant manager.

About Women

In Harold's time there was a commonly held attitude in business and industry toward women. Women were not considered good prospects for advancement in management. If they were married, it was assumed that their obligations to the family would interfere with their performance in the workplace. If they were not married, it was thought that it was only a matter of time before they would marry and start a family and drop out of the workforce. A considerable amount of time and expense was invested in employee training and career development. This was not to be spent on those who were not likely to stay with the company on a long-term basis. This practice was observed at Meriden and long predated the advancement of Harold to his position of leadership in the company.

When the war broke out and John left for military duty, Harold hired Kay Peddle, who was John's cousin, to work as his replacement. She stayed on after the war when he returned. Her responsibilities included being the contact person with the General Electric turbine plant in Schenectady, an important responsibility. In the course of this work she met a GE employee who was a widower with two minor children, one of whom had developmental difficulties. Peddle left Meriden, married the widower and became stepmother to his children. While her dedication was admired, it only reinforced the feeling that women were not interested in career employment. It was a good twenty years before another woman was hired in an executive capacity.

In the sixties and seventies as women came more prominently into the business and professional world, there was mounting pressure to admit women to traditionally all-male organizations. Harold was among those who most strenuously resisted this change. It was his expressed feeling that there should be places where men could be among men. Women had their own social organizations, and men were entitled to equal treatment.

Harold was not a misogynist. There was simply a disconnect between a somewhat dismissive attitude toward women in general and his high regard for women whom he respected and knew personally. Among the many women of accomplishment whom he admired and whose company he enjoyed were Angelina Messina, the editor of *Micropaleontology,* an important scientific journal published by the American Museum of Natural History and printed at Meriden. Catherine Fennelly, the Editor of Publications at Old Sturbridge Village, was a good friend and important customer. Harold was a great supporter and encourager of Gay Walker, then Curator of the Arts of the Book Room at Yale's Sterling Library. Walker became Harold's Boswell and interviewed him extensively regarding his life and work. Her perceptive appreciation of Harold's life and accomplishment was given prominent place at his memorial service at Yale. For all this, it was only after Harold's death that she became the first woman elected to membership in the Columbiad Club.

Seven

A Non-Retirement With Honors

Harold turned the active management of the company over to John in 1975 when he reached the age of 65. He took considerable satisfaction that, having paid into Social Security for many years, he was now about to get some of it back. His reduced responsibilities did not initially involve any significant change in his lifestyle. There was no better life than life at Meriden Gravure and hence no incentive to change his routines. He still had his office, still was in daily contact with the managerial and shop employees, and still greeted the stream of visitors to the shop. He still enjoyed lunch and a game of cribbage at the Home Club. He still went to Boston to the Club of Odd Volumes, usually with Glick as designated driver. He continued to attend the Grolier Club annual meeting and other events in New York during Bibliography Week, and to go to dinners at the Century Association. All this went without having the burden of the day-to-day management of the company.

The honors accorded him began to come even before his retirement. In 1963, Yale awarded him an honorary Master of Arts degree. This was largely through the influence of Wilmarth Lewis, at that time Senior Fellow of the Yale Corporation, although Harold's many other friends in the Yale community certainly joined in the satisfaction of this recognition. The citation read, "In an age which has slight regard for true craftsmanship, you have succeeded in achieving the highest goal of the artist by holding up to nature a perfect mirror. As a coadjutor of writers and scholars, you have illuminated print with faultless taste and added a visual dimension to learning. Recognizing your place in the long and noble New England tradition of creative artisans, Yale confers upon you the degree of Master of Arts."

Seven years later, Wesleyan University bestowed on him an honorary Doctor of Humanities degree. In doing so it recognized his generosity and his many services to Wesleyan as well as his contributions to the scholarly community at large. Wesleyan's honorary doctorate citation read in part, "In an age accustomed to the shoddy and the deliberately obsolescent, you stand as one of those rare preservers of a tradition of excellence. You and your firm are renowned in the graphic arts for impeccable standards of craftsmanship that help us take continuing delight in 'the good, the true and the beautiful.'"

Fig. 7.1. Honorary Degree Recipients at the 1963 Yale Commencement: Front, left to right, Charles D. Dickey; General Lauris Norstad; Yale Provost Kingman Brewster, who bestowed the honors; Morris Hadley; Robert McAllister. Second row, left to right: Harold Hugo; The Rev. Cameron Parker Hall; Herbert Alexander Simon; John Robinson Pierce; Marjorie Hope Nicolson; Samuel Flagg Bemis. (Hugo Family Papers)

Project Viking

In 1977, work began on a festschrift to honor Harold on his retirement. The plan originated with Tom Adams and was quickly embraced by others. A committee was formed which was made up of Julian Boyd, Lyman Butterfield, John Peckham, Marvin Sadik, Walter Whitehill and Rocky Stinehour with Adams as Secretary and coordinator. There was a considerable precedent for such a proj-

Fig. 7.2. 1970 Wesleyan University Commencement: Acting Wesleyan President Robert Rosenbaum, Honorary Degree Recipients Robert Penn Warren, President Theodore Lockwood of Trinity College, Roger Tory Peterson, and Harold Hugo. (Photo by William Van Saun. Wesleyan University Archives)

ect. Various other figures in the scholarly publishing field had been so honored. In all cases the presentation of the publication was an excuse for a good party and was a surprise to the recipient. Fred Anthoensen had been honored in 1952 in a volume entitled *In Tribute to Fred Anthoensen Master Printer.* In 1954 Harold was honored with a keepsake, *Thomas Jefferson Among The Antiquities of Southern France in 1787*, edited by Julian Boyd. The Princeton University Press printed the text, and the collotype illustrations were printed at Meriden without Harold's knowledge. It was presented to Harold at a party in his honor hosted by Walter Whitehill in North Andover. In 1958 Whitehill's many contributions to a wide variety of causes was recognized in *Walter Muir Whitehill:*

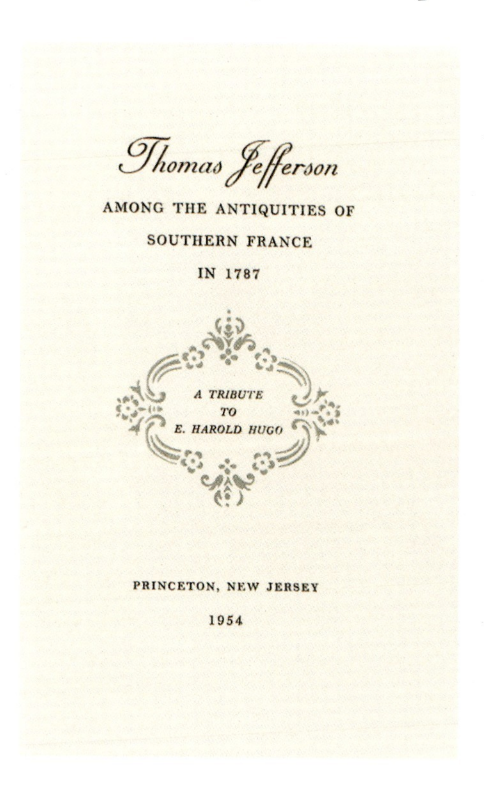

FIG. 7.3. The first of many tributes to Harold. Designed by P. J. Conkwright and printed at the Princeton University Press. (Hugo Family Papers)

A Record Compiled by His Friends. In the following year it was Lyman Butterfield's turn to receive honors when *Butterfield in Holland* was presented to him. With the exception of *Jefferson Among The Antiquities*, Harold was a leading conspirator and contributor on all these publications.

Project Viking was much more than a keepsake. Thirty-six organizations, museums, libraries, research institutions, publishers and bibliophilic organizations participated. Each contributed a fascicle devoted to a treasure in its collection. Each fascicle included a text written by one of the senior members of the organization. Sinclair Hitchings did double duty, writing the text for both the Boston Public Library and the Club of Odd Volumes contributions.

The Stinehour Press was a key participant. It did the typesetting from designs by P. J. Conkwright, the letterpress printing, and it supervised the binding by Coman & Southworth. The Nordic decoration on the title page was Stephen Harvard's rendition of the design that John Howard Benson had cut into the stone that marks the Hugo family gravesite. Each fascicle included illustrations laid in, and these were printed by Meriden, from the originals, of course. This was no easy matter because the illustrations had to be done behind Harold's

A PORTFOLIO HONORING

HAROLD HUGO

FOR HIS CONTRIBUTION TO

SCHOLARLY PRINTING

1978

FIG. 7.4. Title page for "Project Viking."

Fig. 7.5. Hugo Family gravesite marker, Walnut Grove Cemetery, Meriden. Cut in granite by John Howard Benson. (Photo, author)

back and without his knowing. His awareness of what was going through the shop was not greatly diminished by his retirement. The project was given the code name Viking, a reference to Harold's Scandinavian ancestry.

The texts submitted by the institutions varied in character. Some were extremely scholarly and carried footnotes and references. Others were more informal. Many took the opportunity to thank Harold for his advice and service to their institution's publishing program.

Philip Hofer wrote the entry for the Department of Printing and Graphic Arts in the Houghton Library at Harvard.[1] His subject was the Goltzius engraved portrait of Christopher Plantin. It was clearly appropriate that the image of the great sixteenth-century printer was chosen to honor a great twentieth-century printer. Further, it should be mentioned that the print

Fig. 7.6. Hendrik Goltzius, *Portrait Engraving of Christopher Plantin.* First state, before the dedicatory inscription was added at the bottom. (Department of Printing & Graphic Arts, Houghton Library, Harvard University)

was given to The Houghton Library, Harvard, "in honor of E. Harold Hugo from his devoted friend Philip Hofer."

The Dartmouth College Library contribution was a type design in the form of a broadside by Rudolph Ruzicka.[2] It was done in the sixties but did not make the final cut for inclusion in *Studies in Type Design*, published by Dartmouth in 1968 (one of the publications discussed in *Adventures in Printing*). Prompted by Ed Lathem, Ruzicka agreed to let this design be

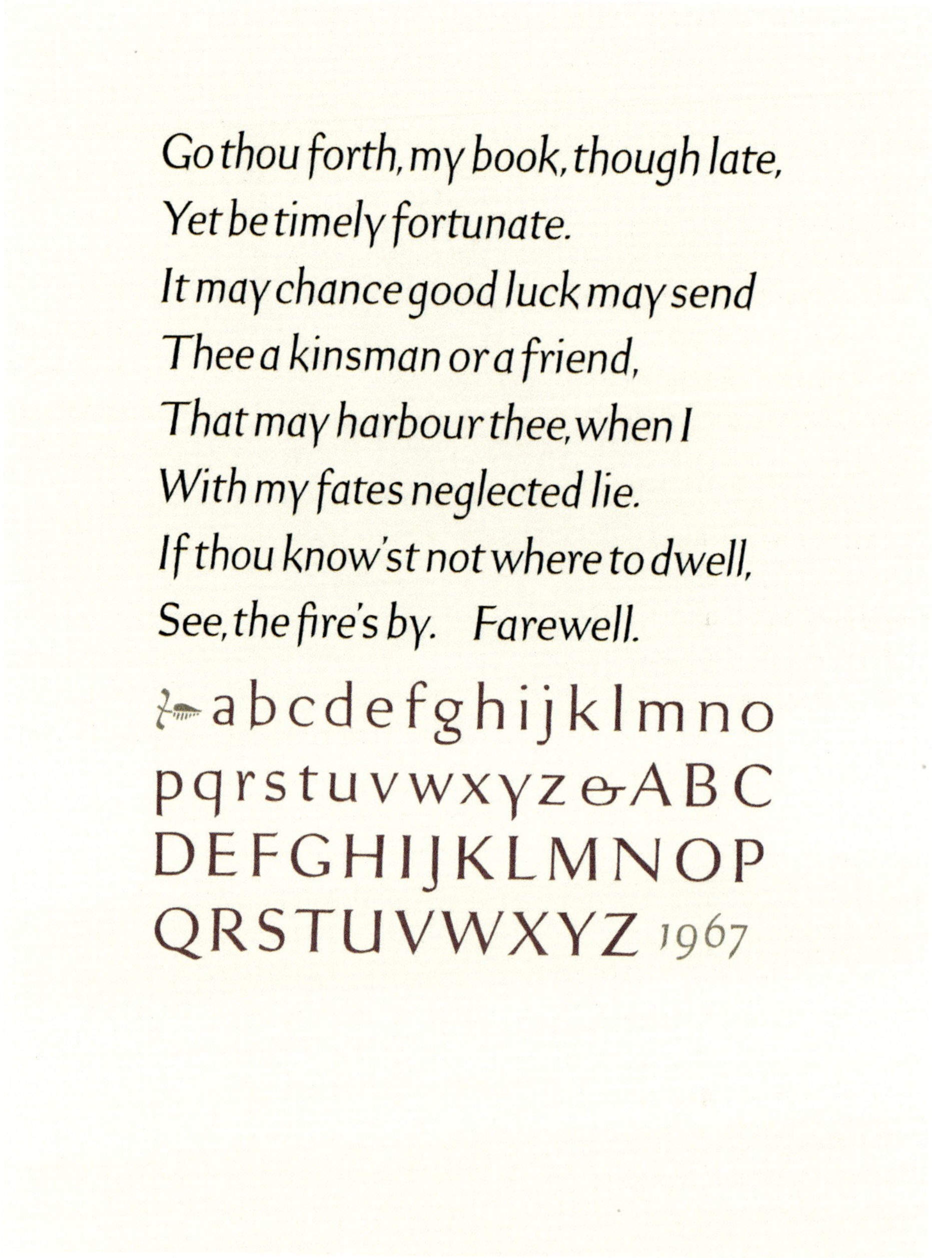

Go thou forth, my book, though late,
Yet be timely fortunate.
It may chance good luck may send
Thee a kinsman or a friend,
That may harbour thee, when I
With my fates neglected lie.
If thou know'st not where to dwell,
See, the fire's by. Farewell.

a b c d e f g h i j k l m n o
p q r s t u v w x y z & A B C
D E F G H I J K L M N O P
Q R S T U V W X Y Z 1967

FIG. 7.7. The "Hugo" type design by Rudolph Ruzicka. The text, "To His Book," is by the Elizabethan poet Robert Herrick. (Dartmouth College Library. Copyright © 1978, Trustees of Dartmouth College)

known as the Hugo letterform. It was a recognition of Rudolph's long and affectionate friendship for Harold. By extension it recognized Harold's long and active support of the Dartmouth College Library. Ruzicka died in July 1978 and so did not live to see the publication of the portfolio three months later.

Carl Zahn was the spokesman for Harold's many friends at the MFA, Boston. In 1969, the MFA published an exhibition catalogue, *Rembrandt: Experimental Etcher*, to mark the artist's death three hundred years earlier. It was one of the highlights of Meriden's long and distinguished relationship with the Prints and Drawings Department. Carl cleverly titled the MFA entry "Harold: Experimental Lithographer"[3] and used the Rembrandt catalogue as a starting point for his account of Harold's contributions to the advancement of offset lithography for scholarly purposes.

Yale was well represented: The Beinecke Rare Book & Manuscript Library, The Yale Center for British Art, The Yale Edition of Horace Walpole's Correspondence and The Yale University Art Gallery were all contributors. Wilmarth Lewis wrote the Walpole entry on Edward Edwards's drawing of The Printing House at Strawberry Hill,[4] and Caroline Rollins wrote a thoughtful and affectionate appreciation of Harold as the Art Gallery entry.[5]

An overseas contribution was received from The Oxford University Press, in which Vivian Ridler, Printer to The University, used a sixteenth-century drawing that Harold had helped him with to express his friendship and admiration for Harold.[6] The Trustees of the Chatsworth Settlement entry was written by Tom Wragg, its Librarian and Keeper. Wragg had known Harold from the sixties when Meriden had printed several exhibition catalogues of old master drawings from the collections of the Dukes of Devonshire that toured the U.S. through the agency of Annemarie Pope's International Exhibitions Foundation.

In addition to the thirty-six fascicles, *A Portfolio Honoring Harold Hugo For His Contributions to Scholarly Printing* contained a number of other important elements. David McCord composed an eight-line poème d'occasion in Harold's praise. Walter Whitehill wrote an informative foreword. Julian Boyd wrote an eloquent and perceptive epilogue. Whitehill and Boyd had not been on the best of terms, a matter of concern to their mutual friends including Adams and Harold. Adams engaged each to play an important part in Project Viking, and in doing so brought about a settling of their differences. As Nicolas Barker wrote in his review of the Portfolio, "it is agreeable to remember that Harold, the great resolver of problems in print should be the cause of their reconciliation."[7]

"Harold At Meriden" by John Peckham added an insider's view of Harold and the company he led. Several tape-recorded interviews dating from 1964 were organized and edited for publication by Sinclair Hitchings. "Harold Hugo on the Meriden Gravure Company" covered aspects of Harold's life and work in his own words. Taken altogether, these elements, now preserved in print, yielded much insight into Harold's personality, his life and his work. Finally, it must be said that the full title of that publication became the inspiration and justification for the title of this present work.

The first copy of the *Portfolio* was presented to Harold at a ceremony held on October 27, 1978, at the Beinecke Library. The wall of secrecy had not been breeched, and the presentation to Harold came as a complete surprise to him. He was greatly touched by this, but perhaps a little shaken that so much of this work had gone on right under his nose, without his being aware of it.

A further honor was the bestowal of the William A. Dwiggins Award by the Bookbuilders of Boston in 1984. Later in the same year John spoke to the monthly meeting of the Club of Odd Volumes on some of Harold's particularly noteworthy efforts in his illustrated talk, "Adventures in Printing."

Bookman and Benefactor

Harold's book-collecting interests centered on important examples of the printer's art over its 500-year history. It was not limited to any narrow time frame or any particular subject area. Incunables were of interest to him. The priciest of these were beyond his reach but, through his friendships with dealers, he had the opportunity and stimulus to purchase many attractive titles at affordable prices. When he did, he would proudly announce his new acquisition to his friend Frederick R. Goff at the Library of Congress, compiler of *Incunabula in American Libraries, A Third Census.* He also acquired highlights from later centuries including Baskerville, Franklin, Updike and others. Harold enjoyed his books, some of which he kept at home and others in his office at work. But he was always willing to part with them if by doing so he could fill in a gap in an important institutional library. In a 1979 letter to Baskin he wrote, "I get as much of a boot out of giving books away as I did in acquiring them." [8] At this time in his life he was doing more giving than acquiring.

Joe Reed wrote of him, "Harold was not only a maker of fine books and

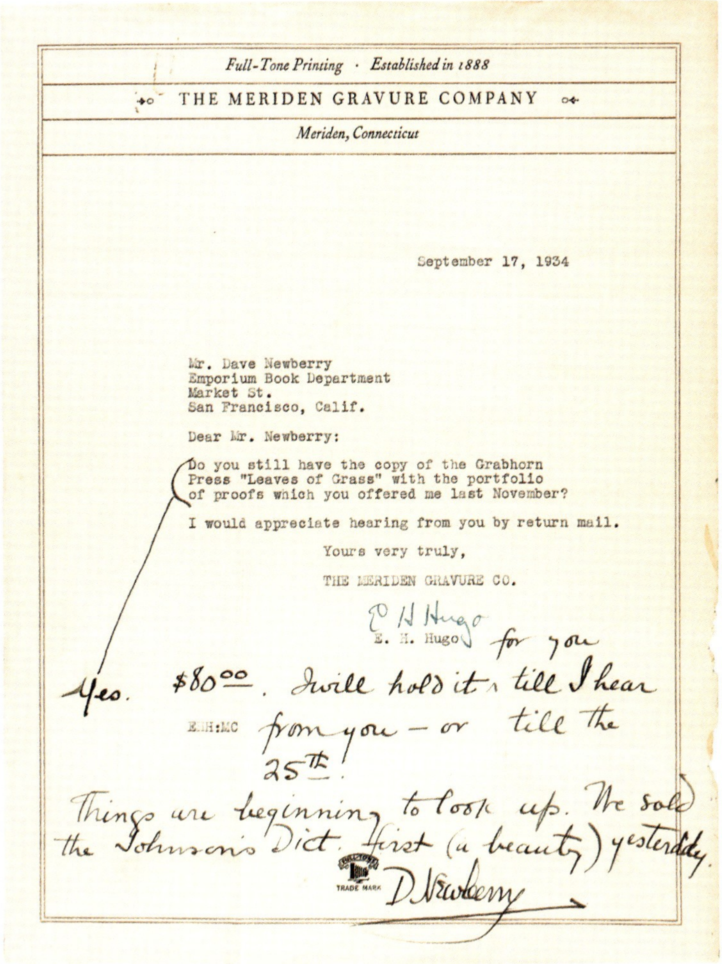

Full-Tone Printing · Established in 1888

THE MERIDEN GRAVURE COMPANY

Meriden, Connecticut

September 17, 1934

Mr. Dave Newberry
Emporium Book Department
Market St.
San Francisco, Calif.

Dear Mr. Newberry:

Do you still have the copy of the Grabhorn Press "Leaves of Grass" with the portfolio of proofs which you offered me last November?

I would appreciate hearing from you by return mail.

Yours very truly,

THE MERIDEN GRAVURE CO.

E H Hugo

E. H. Hugo

E.H:MC

Yes. $80.00. I will hold it for you till I hear from you – or till the 25th!

Things are beginning to look up. We sold the Johnson's Dict. first (a beauty) yesterday.

D Newberry

TRADE MARK

FIG. 7.8. Harold's book order and Dave Newberry's handwritten reply. The Grabhorn *Leaves of Grass* was among the books he gave to Wesleyan University. (Hugo Family Papers)

a perfecter of the printed image, but a custodian of the best that his trade had set forth over all its centuries. He liked to show off a good page, title or colophon or chosen at random in *medias res*. It was clear that he took these objects as the measure for what he wanted to make of books produced in his shop, but equally clear that he loved these works of other great printers for themselves as objects."[9]

He made many gifts during his lifetime and provided for further bequests upon his death. His gifts to Wesleyan University included Benjamin Franklin's *Cato Major*, Updike's *Book of Common Prayer*, the Grabhorn edition of *Leaves of Grass* and Rudolf Koch's *Das Blumenbuch*, given in honor of Wyman Parker on his retirement as University Librarian.[10] The three American books were chosen by Joseph Blumenthal in his *Printed Book in America* as outstanding examples of American printing and were each given a full-page illustration. The Koch book is regarded as one of the masterpieces of twentieth-century German graphic art.

His gifts to Dartmouth included Alcinous's *Disciplinarium Platonis Epitome*, Nuremburg, 1472, from the press of Anton Koberger (Goff A-365), rebound by Arno Werner and given in honor of Ray Nash. He also gave several examples of eighteenth-century New England printing including *The Perpetual Laws*, Worcester, 1788, the work of Isaiah Thomas; the Limited Edition Club 1934 edition of Longus's *Daphnis and Chloe* designed and printed by Porter Garnett at the Laboratory Press; his leaf from the Gutenburg Bible; letters and memorabilia of Rudolph Ruzicka and a collection of Columbiad Keepsakes. Harold was guided in his choice of gifts to Dartmouth by Edward Connery Lathem, Librarian of Dartmouth College, a close associate of Rocky Stinehour and a good friend of Harold as well.[11]

His generosity to Dartmouth included a considerable number of private press books and graphic arts books. Among them were the Grolier Club edition of Dürer's *Of the Just Shaping of Letters* designed by Bruce Rogers, Hermann Zapf's *Manuale Typographicum* and *Typographical Variations* and publications by Devinne, Updike, Anthoensen and Grabhorn that he knew would be useful to Nash in his course work.

Harold's gifts to the American Antiquarian Society were largely eighteenth- and early nineteenth-century imprints that reflected American culture of that time.[12] He gave them his copy of Franklin's 1745 *Westminster Confession of Faith.* Harold was not much interested in theology or liturgy; he probably bought this out of his interest in Franklin and the book's distinguished provenance. A previous owner was the important nineteenth-century collector George Brinley. Several other titles were Hartford imprints. One of these was the first volume (of two) of Bernard Romans's *Annals of the Troubles in the Netherlands . . . A Proper and Seasonable Mirror for the Present Americans.* This was published by Watson and Goodwin in 1778 and dedicated to the patriot governor of Connecticut, Jonathan Trumbull. It was among the books mentioned by Albert Carlos Bates in *Early Connecticut Printing.* Bates says of it, "it is fairly scarce but not a rare book. It is, by the way, the first book in Hartford, and perhaps in Connecticut, to have a rubricated title."[13]

Harold's interest in the AAS inspired Rocky Stinehour to establish a book acquisition fund in his memory. In 1986, one hundred forty of Harold's friends and colleagues contributed $24,000 to the fund. By 2015 the endowment had grown to about $84,000 through additional contributions and through appreciation. Each year several thousand dollars in income from the fund is available to support acquisitions and other activities of the AAS.[14]

Harold gave his copy of *De Chorographia* by the first-century Roman geographer, Pomponius Mela to the John Carter Brown Library in 1969. It was printed in Venice in 1493 by Christoforo de Pensis de Mandello and is listed in the Goff census as M-453.[15] Harold's copy of Dard Hunter's *Papermaking by Hand in America* went to The Massachusetts Historical Society.

Harold's gifts of books to Yale far exceed his gifts to the other institutions here mentioned. Approximately five hundred titles once on his shelves are now at Yale. For the most part they are divided between the Beinecke Library and the Haas Art Library, which includes the Arts of the Book Collection. Included are forty-five incunables and books from every century thereafter. Just as Ed Lathem guided Harold's gifts to Dartmouth, Gay Walker, Curator of the Arts of the Book Collection, and Beinecke librarians, most notably Kenneth Nesheim and Thomas E. Marston, influenced the gifts to Yale. In *The Beinecke Library of Yale University*, published in 2003 to celebrate its fortieth anniversary, Harold is listed in the distinguished company of Louis Rabinowitz, Frank Altschule and Edwin J. Beinecke as important contributors to the Beinecke Library's holdings in early printed books.[16]

Notable books from Harold's collection now at the Beinecke include the following:

Herodotus, *Historiarum libb. LX*, interprete Laurentio Valla, Rome (Arnold Pannartz), 1475, with the bookplates of Juan Moncada y Centellas and Harold Hugo.

Formularium Instrumentorum ad usam Curiae Romanae, Rome c. 1480 with a presentation inscription from Henry Howard, Duke of Norfolk (1628–1684) to the Royal Society of London. The Howards were a leading Catholic family in Tudor and Stuart England.

Guillermus Parisiensis, *Postilla super Epistolas et Evangelia*, Augsburg, 1495, with the bookplates of Harrison Gray Otis (1765–1848) and Harold Hugo, and a presentation inscription to Hugo from Leonard Baskin.

Theocritus, *Eidyllia*, Venice, (Aldus Manutius), 1495/96, with bookplates of Henry Munster and Harold Hugo.

Dante Alighieri, *La Commedia*, with commentary by Cristofaro Landino, Venice, 1497, Petrus de Quarengiis.

Aulus Gellius, *Noctes Atticae,* Paris, 1508, with the bookplates of Harold Hugo, Syston Park and Sir John Hayford Thorold and the autograph of Sir Sidney Carlyle Cockerell, 1901. Cockerell was a close associate of William Morris and later Director of the Fitzwilliam Museum, Cambridge University.

Gabriele Simeoni, *La Vita et Metamorfoseo d'Ovidio,* Lyon, (Jean de Tournes. Typographo Regio), 1584, with bookplates of Horace Goldic and Harold Hugo and an autograph of Matteo Cimini.

Tacitus, *Works*, Antwerp (Officina Plantiniana), 1607.[17]

Many of the above books were given in memory of Carl Rollins. The 1497 Dante was given in honor of Herman W. Liebert. A sixteenth-century Greek New Testament by Estienne was given in memory of Philip Hofer.

The twentieth-century books came from a variety of sources. Yale received Harold's personal copies of many of Meriden's important scholarly books. Others were books Harold bought from dealers. Still others were purchased directly from the private presses that he patronized. He had quite a number of books from the Gehenna Press, as well as Michael McCurdy's Penmaen Press, Carol Blinn's Warwick Press, Clare van Vliet's Janus Press and the Godine Press. From England he had the work of the Golden Cockerell Press, the Nonesuch Press, the Gregynog Press, the Shakespeare Head Press and the Rampant Lions Press. Continental fine printing was represented by the Bremer Presse and the Officina Bodoni. The Arts of the Book Collection also received a considerable quantity of printer's ephemera, some of it printed at Meriden, but much of it printed elsewhere by printers, publishers, artists and dealers who were among Harold's friends and associates.

Harold was always generous in providing his friends with Meriden books. His practice of sending away visitors to Meriden with armfuls of books from the conference room shelves has already been remarked upon. His friends in turn were generous in giving their books to him, and many of his fine press books were gifts to him rather than his purchases. These books are a testimony to his broad interests and his wide circle of devoted friends and associates.

Although most of Harold's older books went to institutions, The Veatchs Arts of the Book dealership, then in Northampton Massachusetts, purchased a number of his books from family members. These books were for the most part his twentieth-century fine press books. Thirty years after his death, some are still available in the rare book market.

A Collector of Art and Artists

In addition to collecting books, Harold collected art. His artistic taste was rather broad. Most of his art was contemporary art and was the work of artists he knew personally. Not surprisingly, his acquisitions were largely works of art on paper: prints and drawings with a few small sculptures, but very little in the way of paintings. The strongest representation in his art collection was that of Leonard Baskin.

In addition to Baskin's strong portraits and expressionistic representations of Everyman in the Atomic Age, Harold enjoyed the work of more relaxed, conservative and representational artists, such as Philip Kappel, Samuel Chamberlain, Francis Comstock, Thomas Nason and Frederick G. Hall, all of whom he knew personally. Kappel depicted the atmosphere of various localities such as Louisiana, Jamaica and New England in books printed at Meriden and published by Little, Brown & Co. He also produced etchings of the same character. Kappel lived in Roxbury, Connecticut. Harold was a frequent luncheon guest there, as was Kappel at Meriden. Sam Chamberlain was often a dinner companion at the Club of Odd Volumes. Chamberlain, Hall, Nason and Comstock were all fascinated with architecture. This was their subject matter, but each had his own distinctive and unmistakable style of representation. Through the energy and initiative of Sinclair Hitchings, Keeper of Prints at the Boston Public Library, Meriden printed publications devoted to each of these artists, and Harold owned original signed prints of each.

The above-mentioned artists were for the most part resident in New England, which facilitated social intercourse on Harold's part. But Harold was also very much attracted to the British School of wood engravers. These included George Mackley, Monica Poole, and calligraphers Reynolds Stone and Leo Wyatt. Wood engraving lent itself to book illustration, and Harold had both books and individual prints representing the art.

Harold did not collect the work of Abstract artists or Pop Art practitioners, although these styles were at their apogee during his lifetime. One small exception to this was Bernard Childs, a New York artist with whom he became acquainted. His color etching in an abstract style hung on Harold's office wall for many years.

Many institutions were favored with his gifts of art during his lifetime, but the principal beneficiary was the William Benton Museum of Art at The University of Connecticut at Storrs. He made gifts during his lifetime. Following his death, his children, Nancy and Gregg Hugo, made further gifts. The founding director of the Museum was Marvin Sadik, whom Harold had met when

FIG. 7.9A. The view out of Harold's office window was no more than a narrow alley leading from the street to the employees' entrance. The visual attraction was entirely within. Hanging on his wall at the upper left is Baskin's woodcut "Self-portrait at 51" and beneath it a lithograph by Stow Wengenroth. On the back wall is Baskin's portrait of Goya and an unidentified relief sculpture, perhaps also by Baskin. To the right is George Grosz's drawing "Two Nudes," now at the Davison Art Center, Wesleyan University. Taped to the glass front of the bookcase are two unidentifiable photos, one above the other. To their right are pictures of Walter Whitehill, a small one of Rudolph Ruzicka as a child, Ruzicka at his worktable, and Chiang Yee in academic gowns. Beneath are pictures of Rollins at his desk, Anthoensen with a book in his lap (see Fig. 6.3) and D. B. Updike in profile. At the bottom is the Bisbee Press, which became the Stinehour Press. (Photo, Robert Hennessey)

FIG. 7.9B. On the wall is Thomas Cornell's drawing of Baskin at Little Deer Isle, now at the William Benton Art Museum, University of Connecticut; a Carolingian manuscript page bought from George Goodspeed; and a color mezzotint by Mario Avati. On top of the left bookcase are three unidentified pieces of sculpture and Baskin's "Owl on a Pomegranate." The photos on the top of the right bookcase are Gregg Hugo's high school graduation picture, Parker in uniform (see. Fig. 3.1), Nancy and Gregg Hugo, and a Rollins broadside. Beneath it are Harold and Rocky, a 1973 photo of Harold in his office when the décor was slightly different (see Fig 5.1), and a photo of Parker at Little Deer Isle with one of the Baskin children, all taken by Tony King. Beneath that, a picture of Julian Boyd, dignitaries at the opening of the Christian Gullager exhibition at the National Portrait Gallery, and the picture of Gregg Hugo in his office (see Fig. 6.18), an unidentified group photo and Ann Stinehour Bottoms with her child. Other children are not identifiable. (Photo, Robert Hennessey)

Sadik was taking the museum course as a graduate student in Fine Arts at Harvard. Meriden printed several of the early exhibition catalogues issued by the museum, which were designed on Sadik's commission by Leonard Baskin. As a newly established museum, trying to build its collections for teaching and exhibition, The Benton was glad to have Harold's support and encouragement. This support continued with Paul Rovetti after Sadik left Storrs for responsibilities elsewhere in the museum world. In 1989 the Museum mounted an exhibition of Harold's gifts and bequests, curated by Thomas P. Bruhn. In the catalogue accompanying the exhibition Stephen Stinehour perceptively noted, "Harold enjoyed collecting, yet he did not collect for himself only. He had a clear vision as to the destiny of his collection and had identified the institutions which he wanted to receive his treasures."[18]

Two important drawings, both of which graced Harold's office, are now at Storrs: Thomas Cornell's drawing *Baskin at Little Deer Isle* and Rico Lebrun's *Three Figures Changed into Snakes*, one of the images inspired by Dante's *Inferno*. Other gifts and bequests that went to Storrs were examples of the work of Connecticut artists Philip Kappel, Thomas Nason, Russell Limbach, Gabor Peterdi and John Taylor Arms.

Important works from Harold's collection also went to the Davison Art Center at Wesleyan, by gift from Nancy and Gregg Hugo. These include the magnificent George Grosz drawing *Two Nudes* and Baskin's large woodcut self-portrait of 1973, both of which had also been in his office for many years. In addition, the DAC received prints by Thomas Nason, Swain Gifford, Louis Novak, Stow Wengenroth, Fritz Eichenberg and other work by Leonard Baskin.[19]

The Twilight Years

Pictures of Harold in his forties show a rather trim figure. But his lifestyle provided ample opportunities to eat, and eat well. As a result, he gained in girth as he grew older. Also, he quit his smoking habit. Tobacco, for all its evils, is an appetite suppressant, and giving up the weed also had an effect on his waistline. The other health problem he had was the loss of one eye when he was in his early fifties as the result of an aneurysm. Carl Rollins lost an eye in his twenties and nevertheless went on to have a distinguished career in the visual arts. Harold must have been aware of this and taken encouragement from it. Observers remarked that Harold saw better with one eye than most others with two. But he had always enjoyed robust good health, and this reminder of his advancing age weighed on him. Never one to complain or feel sorry for himself, he was determined to carry on.

As time went on, Harold's eyesight deteriorated. His eye became tired more readily. He needed a magnifying glass to read small print. A suggestion was made that he might enjoy reading large-print books, but he had no interest in this. He would read his book dealer catalogues and *The Papers of the Bibliographic Society of America* with his magnifying glass and get on as best he could.

In October 1984, Harold experienced dizzy spells and began feeling poorly. He went to his doctor and underwent tests.[20] The diagnosis was poor kidney function, which could be treated only by dialysis. Three times a week, week in, week out, he went to Hartford for treatment. His quality of life was much diminished. His ability to travel was curtailed. He was unable to attend the memorial service for Philip Hofer in Cambridge in November 1984. Later in the same month he was unable to attend the monthly meeting of the COV at which Peckham gave his "Adventures in Printing" talk. The previous monthly meeting in October was the last meeting he is recorded as attending. His attendence at the Columbiad Club meetings was easier to manage, and he attended a majority of them in his last year, including the April meeting, which he moderated. His topic was the work of the English printer Graham Williams of the Florin Press. This press had recently issued a limited edition book of wood engravings by Monica Poole, a favorite artist of Harold. Harold's copy of the book was the centerpiece of his presentation.

In August 1985 a party was arranged to mark his seventy-fifth birthday. The gathering was held in Harold's yard, and the guest list was limited to about a dozen family members and particularly close friends. Photos of the occasion show Harold in good spirits and enjoying the remarks made in his honor. But his cane and the fact that he was so warmly dressed for an August afternoon are evidences of his deteriorating condition. Harold died a month later on September 9, 1985, while undergoing dialysis.

His obituary appeared in the local paper two days later. There was also an article about his life and accomplishments in the news section of the paper. This was a practice of the paper accorded only to Meriden's most prominent citizens. A long and very informative obituary appeared in *The Times* (London) in the September 12 issue. It was unsigned but was unquestionably written by Ruari McLean. It was written well before his death, but in anticipation of it. *The New York Times* dragged its heels. Their obituary was brief and not published until October 5. It was published at that time only after coaxing from John and probably some of Harold's influential friends in New York.

Top Left. John Peckham, Sue Hanna, Harold, Kit Reed.

Top Right. Standing, foreground: John Peckham. Seated: Marge Hugo, Bonnie Hugo, Chao-Wen Chiang, Joe Reed. Standing, rear: Clarence Hugo, Chien-Fei Chiang.

Left. Maria Barnett, Sabrina Stinehour, Steve Stinehour, Nancy Hugo with camera, Jim Barnett.

Fig. 7.10. The 75th birthday party held for Harold in his yard, August 1985. (Photos believed to be taken by Archie Hanna in the Hugo Family Papers)

Harold was laid to rest attended only by his immediate family.

In December there was a memorial service at Yale. Harold had left a note stating that he would enjoy thinking of his friends assembled at a party once more. In response to his wishes, his family and the Yale Library Associates organized a memorial gathering that was the opportunity for those outside the family to pay their respects and bid him farewell. Gay Walker spoke to an audience of 250 at the Yale Law School lecture hall on Harold's life and accomplishments. It was based on her Dwiggins Lecture, "The Meriden Gravure Company & Harold Hugo," given in January 1985 at the Boston Public Library, with modifications appropriate to current circumstances.

Following this, a reception was held in the mezzanine area of the Beinecke Library. While a great many would have liked to say something about Harold, the committee on arrangements wisely limited the speakers to a select five, else the event could have gone on for hours. Peckham served as master of ceremonies and introduced the five, who, he noted, represented but a small sample of Harold's wide circle of friends: a printer (Rocky Stinehour), an historian (Archie Hanna), a university press director (Frank Wardlaw, who unable to attend, submitted remarks to be read by Peckham), a rare book librarian (Tom Adams) and an artist (Leonard Baskin).[21] Each in his own way dwelt on what Harold had meant to them and what he had meant to the world of scholarship and the arts. It was a secular ceremony, a celebration of his life with, appropriately, the magnificent Beinecke stack as a backdrop.

Vision & Revision

INTRODUCING

Meriden-Stinehour Incorporated

MERIDEN · CONNECTICUT AND LUNENBURG · VERMONT

FIG. 8.1. Title page, *Vision & Revision*. Design by Rocky Stinehour and Steve Harvard.

Eight

Changing Times, Challenging Times

As the decade of the seventies began, it became increasingly important for Parker and Harold to consider the future of the company. Meriden had some years previously been approached by several parties looking to acquire it. The most attractive of these offers, according to Harold, was one from the J. Walter Thompson advertising agency. But at the time Parker was not interested in selling and nothing came of the proposal. But now Parker was in his late seventies and widowed. He had previously divested himself of his interests in the Charles Parker Co., which until the Great Depression had manufactured the world-famous Parker shotgun but which in the postwar period had retreated to its more traditional business of turning out kitchen and bathroom fixtures. His children were grown, and neither had the interest or experience to play an active role in managing the company. Gordon, the eldest, was married and living in Florida. His interests were in aviation. He held a small-craft pilot's license, and the Sun Belt climate allowed him to fly year round. Mary Allen Summons, Gordon's younger sister, was married with small children and living in Arizona.

The Merger with Stinehour Press

Parker felt the need to divest himself of his ownership in the Gravure to simplify his estate and provide for his children. But it had to be done in a way that continued the company as a leader in quality illustration printing for scholarly publishers, and that continued to provide local employment for its loyal and skilled employees. It is not clear how many options were available to Harold and Parker at this point. But the one most obvious to them and most comfortable to them was a merger with the Stinehour Press. Stinehour had design, typesetting, letterpress printing and binding facilities that Meriden called upon constantly. Meriden had skilled camera and offset press capabilities that Stinehour required. Harold had known and worked with Rocky for twenty-five years. He was convinced that Rocky shared his commitment to the quality of the product and the satisfaction of the customer. Neither of them could be seduced by the cult of bigness. The working relationship between Meriden and Stinehour had grown over time to the point where each was the other's best customer. They were both about the same size in terms of payroll, revenues and other financial measures.

Stinehour was asked to join the Board of Directors of The Meriden Gravure Co. in 1973.[1] This would clearly indicate that Meriden's Board considered him the most attractive and desired possibility as a merger partner with Meriden. Rumors of the merger were current in 1975, and it may be that an agreement in principle was in effect at that time. But it took another year for the lawyers, bankers and accountants representing the two parties to work out the details of an agreement that would be satisfactory to all concerned. By this time Harold had turned the presidency of the company over to John, but Harold was still very much a presence at 47 Billard St. Harold's help in negotiating the details of the merger agreement was his last great service to Parker and to the company to which he had devoted his life.

The agreement was that a new entity, The Meriden-Stinehour Inc., would over time buy out The Meriden Gravure Company. During that time, Parker, Harold, John and Gordon would have representation on the Board of Directors of the merged company. The company would be incorporated legally and financially in Vermont. Meriden would continue as an autonomous division pretty much as it always had. John, as President of the Meriden Division, would continue to run things in Connecticut. Although it was clear that Stinehour would be the surviving partner in this arrangement, Rocky tactfully insisted that Meriden, as the senior partner in terms of age, be given precedence in the new corporate name. Harold was to be the Chairman of the Board. The merger went into effect in January 1977. When the merger was made public, it was remarked that since Meriden and Stinehour had been living together for so long, it was high time they got married.

Meriden-Stinehour Inc. started 1977 on a strong note. The Bicentennial publication frenzy had subsided, but there was a strong flow of museum publications, university press publications and periodicals to sustain the two divisions. Day-to-day life in Meriden and Lunenburg did not change that much, but there was a considerable increase in phone traffic between area codes 203 and 802. There was also a greater frequency of face-to-face contact between the members of one division and their counterparts in the other. Some of these conferences took place in Connecticut and some in the North Country. One meeting point halfway between Meriden and Lunenburg was McDonald's on fast food alley in Brattleboro. When a more upscale midway gathering place was appropriate, it was often the Putney Inn.

Vision & Revision

To promote awareness and understanding of the capabilities of the merged company to its public, a handsome publication, *Vision & Revision: Introducing Meriden-Stinehour Incorporated*, was produced. This was the inspiration and creation of Rocky and Steve Harvard with Harold contributing to the text. The title was chosen to reflect the past accomplishments of each division and the strength and potential of the merged company going forward. It featured a selection of sensitive photographs by Tony King, showing the Meriden and Lunenburg workplaces and the men and women who labored there. Pictures of the two plants were intermixed without identification of location, but the people shown were identified in the combined list of employees. The clear import of this was to stress the contribution of these craftspeople, and the unity and fusion of spirit in the merged company.

Vision & Revision was issued to friends and admirers of the firm in a hardcover edition, bound in the Press's favorite russet Scholco cloth (also used on the *A Portfolio Honoring Harold Hugo*.) But prior to that, in 1979, a paperbound pre-publication edition of one hundred copies was produced to honor Parker's eighty-fourth birthday. This reflected Rocky's high regard for Parker, his wish to honor the grand old man of the merged company, and his desire to offer reassurance that the enterprise to which Parker had devoted so much of his life would be in good hands unto the next generation.

In the following year Parker, the great athlete who, well into his senior years, still took the steep factory steps two at a time, suffered a fatal heart attack and was gone. Although he had long since relinquished his formal duties and responsibilities, he was, to his last days, the engaged observer and active encourager of the work and workers at 47 Billard St. He was mourned and missed not only by the employees, but also by his social friends at the Home Club and by his many associates in the business and civic life of Meriden.

John's experienced hand guided The Meriden Division until his retirement. John's strength was his ability to navigate the familiar waters previously explored and charted by Harold. But he was not inclined to take the risks of voyaging in unfamiliar waters, and in the post-merger period any venturesome initiatives would have to come from Lunenburg anyway. In 1984 he was succeeded by Stephen Stinehour, who was brought into the office to assume managerial and sales duties. Steve, a generation younger and with less experience, nevertheless brought a fresh energy and the entrepreneurial spirit inherited from his father to the leadership of the Meriden Division.

Fig. 8.2. Rocky Stinehour and Steve Stinehour in 1984 at about the time Steve succeeded John Peckham as President of the Meriden Division. (Photo, Alan Rodgers)

The Consolidation

The merger of 1977 envisaged two autonomous divisions. But in the ensuing decade much had changed. The merger had stimulated commerce between the two divisions, which was good. But this had the effect of making the two divisions more interdependent on one another and less autonomous. The development of phototypesetting, already well underway, favored offset presswork to the detriment of letterpress. The decade following the merger was one of tectonic change in the printing industry. Classic typefaces, the mainstays of quality book composition, became available in digital form, and phototypesetting gave way to digital typesetting. Desktop computers became readily available, along with software that combined design, word processing and page makeup with illustrations integrated into the text. No longer was it necessary to go to a typesetting house with a huge investment in machines and matrices to buy composition. Pre-press functions became more decentralized and more under the control of authors, editors, designers and publishers. But "everyone his own editor/designer/compositor" was not without its drawbacks; many who attempted it were in over their heads in these specialized disciplines. The Stinehour Press had an important role to fill in this situation because it had the specialized design and editorial skills that were needed in the brave new digital world. Steve Harvard was a key figure in this transformation. He had a thorough grounding in traditional design and bookmaking. But through his work with Mergenthaler and later with Adobe he understood both the opportunities and the problems in going digital. He took it as his mission to apply the promise of the digital revolution to the craft tradition of fine printing.

All of this meant that design and composition and makeup would be done in Lunenburg, negative assembly, platemaking and presswork in Meriden, and binding and shipping back at Lunenburg. Some binding could have been done elsewhere. The Mueller Trade Bindery in Middletown bound many of Meriden's paperbound publications. Acme Bookbinding in Boston was a competent and reliable source of hardcover binding. But Stinehour had expanded its bindery over time to include both paperback and hardcover binding as well as mailing services for its numerous periodicals. Rocky's vision from very early in his career was to be a full-service book producer with the capability for everything from design, copyediting and composition to binding and final delivery. With this in mind, the organizational and physical back-and-forth between the two locations appeared awkward and inefficient.

A small step was taken in 1985 to partially alleviate this problem.[2] An offset press was installed in Lunenburg to do small, easy offset work, typically illustration inserts, covers and jackets, and periodicals. This required not only press equipment and personnel, but also the pre-press equipment and personnel to support it. But the amount of offset work that could be done at Lunenburg was not of sufficient volume to be efficient. Thus, one set of inefficiencies had been replaced by another, and the problem remained unsolved.

This led to a consideration of a more extensive integration of the two divisions. This would not only bring about efficiencies in the production process, but would also facilitate better managerial oversight. Further, there was some redundancy in office work made necessary by the two-division structure, and this could be eliminated. The cost saving in doing so would be most welcome. Beyond this, there were philosophical and psychological implications. The unified, one company concept that Rocky had been at pains to emphasize in *Vision & Revision* would be fostered and encouraged.

If there were to be a consolidation of facilities, where would it be located? Both Meriden and Lunenburg were given consideration. Meriden's chief advantage would be its greater proximity to the markets that Meriden-Stinehour served. But Stinehour had been successful in generating work in Boston and New York from its Lunenburg location, and had every confidence that it could continue to do so. A number of factors weighed in the balance on the side of a Lunenburg location. Housing costs in Connecticut were higher than in the Lunenburg-Lancaster area. This would work a hardship on families relocating from the North Country to central Connecticut. On the other hand, cheaper housing would be an advantage to Meridenites willing to go north. Because of cost-of-living factors, wage rates in Lunenburg could be a little lower, and this would offer some advantage against national and international competition. Finally, Stinehour was the surviving partner in the merger and Lunenburg was already the legal and financial headquarters of Meriden-Stinehour Inc.

Beyond these business considerations, there were personal issues. Rocky had established himself in the North Country in part due to circumstances, but in greater part by design. The North Country ethos was an integral part of the culture of the Stinehour Press as he fashioned it. Rocky's two principal lieutenants, Steve Harvard and Freeman Keith, were both originally flatlanders who emigrated to the Northland to live and work. They, as well as other local people, had no desire to give up their citizenship in the Northeast Kingdom. As it turned out, neither Freeman nor Steve were with the Press when the consolidation actually went into effect in 1989. Freeman had health problems and had to retire in 1987. Steve Harvard died tragically the following year. But the planning for the consolidation predated these events, and personal feelings on the matter undoubtedly influenced the decision. Taken all together, the balance scales were clearly weighted in favor of a Lunenburg location. And so it came to be.

The announcement of the consolidation was made to Meriden employees in November 1988. The Meriden plant was to be shut down in June of the following year. Any Meriden employees who wished to go north would have jobs available to them at Lunenburg, and it was hoped that many would find this offer attractive. Nine of the thirty-nine employees on the payroll accepted; the rest declined. The resistance to going was in most cases a matter of family considerations. Many had spouses with good, steady jobs in the Meriden area. Others had responsibilities to aged parents, or children settled in a familiar school system, or were parents helping their adult offspring with caring for the grandkids. A few were so close to retirement age that it did not make sense to move. Jobs in the Meriden area were readily available. Other printers in the area were only too happy to pick up Meriden's well-trained shop workers and experienced supervisors. The office staff had little trouble finding jobs for which they were qualified in other manufacturing plants around town. The situation in which Harold's son, Gregg, found himself was fairly typical. He and his wife, Bonnie, had a home adjacent to his parents' home. When Harold died, Gregg and Bonnie took on the responsibility of helping Marge Hugo so that she could continue to live independently in her own home. Gregg got a job with Miller-Johnson Printing Company and remained in Meriden. He could not do otherwise.

If there were any miscalculation about the consolidation on management's part, it was that it overestimated the flexibility and mobility of the Meriden workforce. It proved easier to relocate equipment than to relocate people. But the Stinehour management made do. Those remaining open positions in Lunenburg were filled by local North Country people or by new hires recruited from elsewhere in New England.

When the plant shut down, the best and most useful equipment was shipped up to Lunenburg. Some of the remaining equipment was sold to second-hand machinery brokers. What they would not take went to the scrap dealers. The land and buildings at 47 Billard St. were put up for sale. They were subsequently sold to a wholesale picture-frame business, which imported ready-made frames from overseas, warehoused them in Meriden, and sold them to art supply stores.

The Meriden-Stinehour Press in the post-consolidation period continued to do fine work, but there wasn't much Meriden in Meriden-Stinehour anymore. Before long Meriden was dropped from the name, and The Meriden Gravure Co. faded into history.

Nine

Conclusion: "A Broad And Humanizing Employment"

Printing is called "the art preservative," and Harold was by this definition a dedicated and expert preservationist. He was not himself a producing scholar. He was, rather, midwife to the scholarship and creativity of others. Under his care, their efforts came into the world, not infrequently after a lengthy and difficult period of gestation, successfully delivered, and fully and perfectly formed. Alvin Eisenman, who thought broadly and deeply about the practice of preserving and disseminating knowledge, said of Harold in 1978, "there has never been anyone who held the position that Harold does in American scholarly printing."[1] Gay Walker summed up Harold's gifts very economically when she said in her Dwiggins Lecture, "Harold's trademarks were his foresight and initiative, his amazing capacity to make good, lifelong friends, and his uncompromising eye for top quality work."[2]

To reprise M. E. Brasher: "the compact with Meriden Gravure officers was altogether heartwarming and inspiring. Here were businessmen willing and anxious to help a white-haired man realize a dream . . . It was indeed a tribute to Rex. It was a tribute to his work. There could be no other inference." Made when Harold was but nineteen, it was an early expression of qualities that would characterize his subsequent career. It was only the first of many occasions when the inspired dreams of others were realized by Harold's effort and determination.

Harold was faithful to his convictions; he was a stubborn idealist working in a world he knew perfectly well was far from ideal. He brought out the best in the people around him because he always labored to bring out the best in himself. He was the conscience of his profession. Frank Wardlaw in his memorial remarks said, "I never published a book which I thought was beautiful without hoping instantly that Harold would share my pride in it. And I never published an inferior book—there were plenty of them too—without hoping that Harold would never see it."[3]

Julian Boyd said of him, "We can honor him most by endeavoring in our varied tasks to perpetuate those principles of excellence, of integrity, and of pride of craftsmanship which he has exemplified so steadfastly throughout his entire career."[4]

Updike, in the conclusion of *Printing Types*, wrote,

> For a printer there are two camps, and only two, to be in: one the camp of things as they are; the other, that of things as they should be. The first camp is on a level and extensive plain, and many eminently respectable persons lead lives of comfort therein; the sport is, however, inferior! The other camp is more interesting. Though on an inconvenient hill, it commands a wide view of typography, and in it are the class that help on sound taste in printing, because they are willing to make sacrifices for it. This group is small, accomplishes little comparatively, but has the one saving grace of honest endeavor—*it tries*. . . . You may not make all the money you want, but you will have all you need, and moreover, you will have a tremendously good time; . . . The practice of typography, if it be followed faithfully, is hard work—full of detail, full of petty restrictions, fully of drudgery, and not greatly rewarded as men now count rewards. There are times when we need to bring to it all the history and art and feeling that we can, to make it bearable. But in the light of history, and of art, and of knowledge and of man's achievement, it is as interesting a work as exists—a broad and humanizing employment which can indeed be followed merely as a trade, but which if perfected into an art, or even broadened into a profession, will perpetually open new horizons to our eyes and new opportunities to our hands.[5]

Updike's emphasis was on typography, but if expanded to include the book arts more broadly, it could serve as Harold's credo as well.

Harold's remains are interred in Meriden's Walnut Grove Cemetery. His grave marker is there in the family plot, but his real and lasting monument is his work, the books that he painstakingly brought into being. They survive and will be admired and instructive for generations to come.

Appendix: Books Discussed by John F. Peckham in *Adventures in Printing*

Jefferson, Thomas. *The Declaration of Independence.* Washington: The Library of Congress, 1943.

Kennedy, Ruth. *The Renaissance Painter's Garden.* New York: Oxford U.P., 1948.

Brigham, Clarence. *Paul Revere's Engravings.* Worcester, MA: The American Antiquarian Society, 1954.

Lewis, Wilmarth and Philip Hofer. "*The Beggar's Opera" by Hogarth and Blake.* Cambridge and New Haven: Harvard University Press and Yale University Press, 1965.

Van Nice, Robert. *Santa Sophia in Istanbul.* Washington: Dumbarton Oaks, 1965.

The Artist and the Book 1860–1960. Boston: The Museum of Fine Arts, 1961.

Catherwood, Frederick. *Views of Ancient Monuments in Central America, Chiapas and Yucatan.* Facsimile of the text and 25 plates, originally published in London, c. 1844. Barre, Mass.: Barre Publishers, 1965.

Des Barres, Joseph F. W. *The Atlantic Neptune.* Issued in four series. Barre, MA: Barre Publishers, 1966–1969.

Benson, John Howard. *Operina: The First Writing Book.* New York: The Chiswick Book Shop, 1954.

Comstock, Francis A. *A Gothic Vision: F. L. Griggs and His Work.* Boston: The Boston Public Library, 1966.

Whitehill, Walter M. and Rudolph Ruzicka. *Boston.* Boston: David R. Godine, 1975.

Ruzicka, Rudolph. *Studies in Type Design.* Hanover, NH: Dartmouth College Library, 1968.

Peter, Armistead III. *Tudor Place.* Georgetown, D.C.: privately printed for the author, 1969.

Endnotes

Abbreviations:
HFP: Hugo Family Papers
EHH: Everett Harold Hugo

One: The Early Years of Meriden Gravure, 1888–1923 (pp. 1–8)

1. "Getting no sympathy": Gay Walker, *Harold Hugo & the Meriden Gravure Co.*, n.p. (p. 3).
2. "favorable to the lender": Walker, op. cit., n.p. (p. 3).
3. "Collotype takes its name": Chayt, *Collotype*, p. 2.
4. The collotype process: Stulik and Kaplan, *The Atlas of Analytical Signatures of Photographic Processes: Collotype*, pp. 4–10, 15–17.
5. "Bierstadt did not continue working with Prang": Helena E. Wright, "Bierstadt and the Business of Printmaking," pp. 267–274 in Anderson, Nancy K. and Linda S. Ferber, *Albert Bierstadt: Art & Enterprise.*
6. "making collotype prints derived from his brother's paintings": Kelman, "Fine Press Printing: The Meriden Gravure Co.," *Journal of the New Haven Colony Historical Society*, Volume 31, No. 3, Summer 1985, p. 64.
7. "Allen's employment": Meriden City Directories for 1888–1891.
8. "Allen married Cornelia Parker Breese": Edith Wheeler Robinson, *History of the Thursday Morning Club*, n.p. (Meriden, CT), Privately Printed for the Members, 1941, p. 10.
9. "debt was fully satisfied in 1910": Walker, op. cit., n.p. (p. 3).
10. "giving out tickets": Harold's interview in Epilogue Fascicle in *Portfolio to Honor Harold Hugo*, n.p.
11. "no truth to this": Marazzi, *A Bowl Full of Memories: 100 Years of the Yale Bowl*, New York, 2014, Sports Publishing, p. 18.
12. Bachelor's degree, *honoris causa*: Caryn Spies, Manuscript & Archives, Sterling Memorial Library, Yale University, to author, 11/5/15.

Two: Harold Finds His Calling (pp. 9–27)

1. Birth of EHH: Copy of Certificate of Birth. It erroneously lists his father's name as Arthur rather than Otto. HFP.
2. Move to Meriden: Price & Lee Meriden City Directories for 1917–1921.
3. "Harold a member of the Meriden High School Class of 1927": *Muse: The Meriden High School Yearbook for the Class of 1927*. HFP.
4. "general laborer and not a skilled craftsman": Price & Lee Meriden City Directories for 1917–1928.
5. "Harold learned to drive": John Peckham, "Harold at Meriden," Epilogue Fascicle in *Portfolio to Honor Harold Hugo*, n.p.

6. "Harold did well in his freshman year": Northeastern University report cards in HFP.
7. "Letter of withdrawal": EHH to Dean Carl S. Ell, 10/16/28. HFP.
8. "1931 was bad and 1932 was worse": Kelmen, op. cit., p. 66.
9. "publication work for universities": Walker, op. cit., n.p. (p. 8).
10. *The Permian of Mongolia* by A.W. Grabau": Walker, op. cit., n.p. (p. 8).
11. "reviewed in a geological journal": R. W. Sherlock, *Geological Magazine*, Vol. 68, Issue 07, July 1931, pp. 334–335.
12. "Professor Ray Moore": Walker, op. cit., n.p. (p. 9).
13. Arthur C. Sias, "The Full-Tone Collotype Printed Reproduction," *Journal of the Biological Photographic Association*, Vol. 1, No. 2, December 1932, pp. 82–88.
14. Carl O. Dunbar, *The Illustrations of Paleontologic and Geologic Literature.* Full-Tone Collotype for Scientific Reproduction, Supplement No. 5. Meriden, Conn.: The Meriden Gravure Co., n.d.
15. Milton E. Brasher, *Rex Brasher, Painter of Birds, A Biography.* New York: Rowman & Littlefield, 1961, pp. 264–265.
16. Milton E. Brasher, op. cit., p. 209.
17. "addiction of printing for pleasure": John Peckham, "Harold at Meriden," Epilogue Fascicle in *Portfolio to Honor Harold Hugo*, n.p.
18. EHH to Albert Carlos Bates, 4/18/34. HFP.
19. Edmund Gress, *The American Printer*, July 1934.
20. C. P. Rollins, "The Compleat Collector," *The Saturday Review of Literature*, February 2, 1935.
21.. Distribution list in Harold's hand. HFP.
22. Ward Ritchie letter. Walker, op. cit., n.p. (p. 21).
23.. The Meriden Gravure Co. WW II Honor Roll in the collections of the Meriden Historical Society.
24. "Fry & Kammerer": Blumenthal, *The Printed Book in America*, p. 16–17.

Three: Meriden Reinvents Itself (pp. 28–67)

1. Robinson patent for electroplating gelatin plates: *U.S. Patent Office Publication June 2, 1931*, p. 121.
2. Communication from Michael Frost, Manuscripts and Archives, Sterling Memorial Library, Yale University, April 13, 2016.
3. Communication from James Workman, Center for Technology and Research, Printing Industries of America, Feb. 16, 2016.
4. Clarence Kennedy: David Bourbeau, *Out of the Cellar: A Garland for Cantina*, p. 15.
5. Sachs & Mongan: Kelman, op. cit., p. 67.
6. Ledoux Japanese Prints: Kelman, op. cit., p. 67.
7. Office of Strategic Services: Gay Walker, "Printing for the United States: Meriden Gravure and the U.S. Government," *Printing History* 20, Vol. X, No. 2, p. 22.
8. Gay Walker, "Printing for the United States: Meriden Gravure and the U.S. Government," *Printing History* 20, Vol. X, No. 2, p. 22.

9. "It did not occur to us": Typescript of talk given to the Design and Production Session at the AAUP Annual Meeting, June 18, 1973, p. 4. Meriden Gravure papers at the Meriden Historical Society.
10 "This was in 1957": Carl F. Zahn, Museum of Fine Arts, Boston, *Portfolio to Honor Harold Hugo*, Fascicle 20.
11. "Harold, the trouble with you": Stephen Riley quoted in *Thomas Jefferson Among the Antiquities of Southern France in 1787: A Tribute to E. Harold Hugo*, p. 3.
12. Ridler statement: *Portfolio to Honor Harold Hugo,* Fascicle 26.
13. William Bostick: "Recent Publications in the Field of Art," *Art Quarterly*, Vol. XXVII, No. 4, pp. 529–531. Detroit: Detroit Institute of Art, 1964.
14. "It took only a short time": Archibald Hanna, *Portfolio to Honor Harold Hugo,* Fascicle 34.
15. Naval officer's security concerns: Gay Walker, "Printing for the United States: Meriden Gravure and the U.S. Government," *Printing History* 20, Vol. X, No. 2, p. 22.
16. Chinnery "fakes": Ernest Dodge, *Portfolio to Honor Harold Hugo,* Fascicle 27.
17. Rorimer: *Time Magazine*, January 26, 1962, p. 72

Four: Life In the Age of Offset (pp. 68–98)

1. Duotone negatives: Information from Richard Benson to author, 8/26/14.
2. "Harold was thanked": *A Tribute to Philip May Hamer on the Completion of Ten Years as Executive Director of the National Historical Publications Commission by a Few of the Many that Have Benefitted from His Labors,* New York, December 29, 1960.
3. "If I were asked by a harassed St. Peter": Joseph Blumenthal, *Typographic Years: A Printer's Journey Through a Half Century, 1925–1975*, pp.120–121.
4. Talk to the Philobiblon Club of Philadelphia: Typed copy of the talk in the author's possession.
5. "copies were soon confiscated": Gay Walker, "Printing for the United States: Meriden Gravure and the U.S. Government," *Printing History* 20, Vol. X, No. 2, p. 32.
6. "the notion that incised stones": Carl Schuster and Edmund Carpenter, *Patterns That Connect*, p. 8.

Five: Running the Show (pp. 99–113)

1. Typescript of talk given to the Design and Production Session at the AAUP Annual Meeting, June 18, 1973, p. 3 Meriden Gravure papers at the Meriden Historical Society.
2. E. F. Schumacher, *Small Is Beautiful: Economics as If People Mattered.* London: Blond & Briggs, 1973.
3. P. J. Conkwright, "Types and Time," *Scholarly Publishing*, Vol. 5, No. 2, January 1974, pp. 121–124.

Six: A Man Among Men (pp. 114–144)

1. Bibliographic Press: Katherine McCandless Ruffin, "Carl Rollins and the Bibliographic Press," talk delivered to The Society of Printers, October 1, 2014.
2. "not well informed": Gay Walker, *Harold Hugo & the Meriden Gravure Co.*, n.p. (p. 7).
3. Standard's complaint to Updike: Letter of July 8, 1937, copy to Hugo. HFP.
4. Updike reply to Standard, July 12, 1937, copy to Hugo. HFP.
5. Updike complaint to Hugo, July 12, 1937. HFP.
6. Hugo to Updike, undated but referred to by Updike as being written July 22, 1937, handwritten draft. HFP.
7. Standard's apologetic letter: Updike to Hugo, July 23, 1937. HFP.
8. "strong emotions of Updike and Standard": Standard to Hugo, July 23, 1937. HFP.
9. "Fred Anthoensen decided": Walter Muir Whitehill, *Fred Anthoensen: A Lecture*, New York: The Composing Room, 1966, p. 10–11 (reprinted in *Heritage of the Graphic Arts*, New York: R. R. Bowker, 1972, p. 219).
10. "model of Ned Thompson": Edmund Thompson, "Hawthorn House: A Record of the First Five Years," *The Annual of Bookmaking*, New York: The Colophon, 1938, n.p.
11. *Loyalist Operations at New Haven*: "Notes for Bibliophiles," edited by Lawrence C. Wroth, *New York Herald Tribune* (book section) Sunday, April 24, 1938.
12. Thomas R. Adams as successor to Wroth: *In JCB: An Occasional Newsletter of the John Carter Brown Library*, Harold Hugo Memorial Minute, n.d. (1986).
13. Thomas R. Adams on Hugo contribution to the *Blathwayt Atlas*: Foreword to *Blathwayt Atlas*, Volume I: The Maps, n.p.
14. John Howard Benson's relationship with Hugo: Richard Benson to the author, 5/27/14.
15. Ruari McLean, "The Reproduction of Prints," *Print Quarterly*, Vol. 4, No. 1, London, March 1987, pp. 40–45.
16. *Old Sturbridge Village Research Society Newsletter*, Number 11, February 1986. HFP.
17. Richardson Lake: Ruari Mclean, *True to Type*, pp. 115–119.

Seven: A Non-Retirement With Honors (pp. 145–165)

1. Philip Hofer, *Portfolio to Honor Harold Hugo*, Fascicle 14.
2. Edward Connery Lathem, *Portfolio to Honor Harold Hugo*, Fascicle 10.
3. Carl Zahn, *Portfolio to Honor Harold Hugo*, Fascicle 20.
4. Wilmarth S. Lewis, *Portfolio to Honor Harold Hugo*, Fascicle 36.
5. Caroline Rollins, *Portfolio to Honor Harold Hugo*, Fascicle 33.
6. Vivian Ridler, *Portfolio to Honor Harold Hugo*, Fascicle 26.
7. Nicolas Barker, "News and Comment," *The Book Collector*, Vol. 29, No. 3, Autumn 1980, pp. 419–420.

8. Hugo to Baskin in England, August 15, 1979. Copy in notebook at the Meriden Historical Society.
9. Joseph Reed, Memorial Minute as printed in the *Yale Library Gazette*, Vol. 60, Nos. 3–4, April 1986.
10. Wyman W. Parker, *Portfolio to Honor Harold Hugo*, Fascicle 32.
11. Dartmouth gifts: Information supplied to author by Morgan Swan, Dartmouth College Library, July 1, 2014.
12. AAS gifts: Information derived from online search of the American Antiquarian Society card catalogue (http://catalog.mwa.org/), 7/17/13.
13. Albert Carlos Bates, *Some Notes on Early Connecticut Printing*, pp. 11–12 in Meriden edition of 1934, privately printed by Gregg Anderson and Harold Hugo. Reprint from *The Papers of the Bibliographical Society of America*, Vol. 27, Part 1.
14. AAS Hugo Book Fund: Mary Callahan to Marge Hugo, 1/21/86 (HFP); News-Letter of the AAS, No. 37, February 1986; information to author from Thomas Knoles, AAS, July 2014.
15. Pomponius Mela: Information from Kim Nusco, Reference Librarian, JCB, 2/28/14.
16. Babcock, Robert G., "Early Manuscripts and Books" in *The Beinecke Library of Yale University*, Stephen Parks (Editor). New Haven: Beinecke Rare Book & Manuscript Library, 2003, p. 70.
17. Yale gifts from EHH: Lists and letters of receipt from Yale University Library, various posthumous dates. HFP.
18. Stephen Stinehour quoted in Bruhn, Thomas P., *Harold Hugo (1910–1985), Museum Patron*, Storrs, CT: William Benton Museum of Art, 1989.
19. Gifts to the Davison Art Center: Ellen D'Oench acknowledgment list dated 2/3/87 and letter dated 9/22/89. HFP.
20. EHH declining health: Gregg Hugo, undated oral interview.
21. Copies of memorial remarks by John Peckham, Leonard Baskin, Tom Adams and Frank Wardlaw are in HFP. Rocky Stinehour and Archie Hanna's remarks not available.

Eight: Changing Times, Challenging Times (pp. 166–173)

1. Stinehour on MG board: Gay Walker, *Harold Hugo & the Meriden Gravure Co.*, n.p.
2. "A small step": Stinehour Press Newsletter, Vol. 1, No. 1, 1985.

Nine: Conclusion: "A Broad And Humanizing Employment" (pp. 174–175)

1. Alvin Eisenman: Quoted by Julian Boyd in *Portfolio to Honor Harold Hugo*, Epilogue Fascicle.
2. "Harold's trademarks": Gay Walker, *Harold Hugo & the Meriden Gravure Co.*, n.p (p. 2).

3. Frank Wardlaw: Typescript of remarks written for the Memorial Service at Yale University December 6, 1985. HFP.
4. Julian Boyd, *Portfolio to Honor Harold Hugo,* Epilogue Fascicle.
5. D. B. Updike, *Printing Types: Their History, Form and Use,* 2nd Edition, Cambridge, MA: Harvard University Press, 1937, Vol. II, pp. 275–276.

Sources

Hugo Family Papers (HFP): Harold's business correspondence is a part of the Meriden Gravure Archive at the Beinecke Rare Book & Manuscript Library at Yale. He kept some correspondence of a more personal nature, however, and this has passed down to the family. It includes material relating to Timothy Press publications and correspondence with book dealers. It also includes some childhood memorabilia, and materials relating to his high school and college career, and bequests to various institutions. The Hugo Family Papers, presently in the possession of Harold's son, Gregg Hugo, are destined for the Beinecke Library.

Meriden Historical Society (www.MeridenHistoricalSociety.org): The Morehouse Research Building, 1090 Hanover St., South Meriden 06451, has considerable material on the Meriden Gravure Co.

Meriden Public Library: The Local History Collection has a folder of newspaper clippings related to the company's later history and also a file of Warren F. Gardner's essays and reportage. Gardner was a high school classmate and lifelong friend and admirer of Hugo. In his distinguished career in journalism he served as a reporter, columnist, and essayist. He was for twenty-four years editor of the local morning paper, the *Meriden Record*, and its successor, the *Meriden Record-Journal*. On four significant occasions, listed in the Bibliography, he wrote feature articles about Hugo.

Yale University: The Meriden Gravure Archive is at the Beinecke Library, Yale. It is much more extensive for the postwar period than for earlier years. It occupies approximately 400 linear feet and has the call number GEN MSS 606. The Yale University Library, Manuscripts and Archives, has biographical information on Yale graduates associated with the Meriden Gravure Co. The Robert B. Haas Family Arts Library, which includes the Arts of the Book Collection, has many examples of Meriden Gravure Co. printing.

Annotated Bibliography

Anonymous. "The Meriden Gravure Co." *Centennial History of Meriden.* Meriden, CT: Journal Publishing Co., 1906. Part III, p. 123, with illustration of the plant.

Adams, Thomas B. "Harold Hugo." *In JCB: An Occasional Newsletter of the John Carter Brown Library*, Fall 1985. A memorial account summarizing his life, with particular emphasis on his service to the JCB.

Anderson, Nancy K. and Linda S. Ferber. *Albert Bierstadt: Art & Enterprise.* New York: Hudson Hills Press, 1990. Includes Helena E. Wright, "Bierstadt and the Business of Printmaking," pp. 267–294.

Barker, Nicolas. A review of *A Portfolio Honoring Harold Hugo* and *Vision & Revision: Introducing Meriden-Stinehour Press. The Book Collector*, Volume 29, No. 3, Autumn 1980, pp. 419–420.

Benson, Richard. *The Printed Picture.* New York: The Museum of Modern Art, 2008. A comprehensive history of the technology of illustration printing from the Renaissance to the Digital Age. Chapter 11 covers the practice of collotype and offset over the life span of the Meriden Gravure Co. and includes specific reference to Meriden.

Blumenthal, Joseph. *The Printed Book in America.* Boston: David R. Godine, Publisher, 1977. Paragraph on the Meriden Gravure Co. in chapter on 20th Century New England Printers, p. 127.

Blumenthal, Joseph. *Typographic Years: A Printer's Journey Through a Half Century, 1925–1975.* New York: Frederic C. Beil, 1982.

Bourbeau, David. *Out of the Cellar: A Garland for Cantina.* Northampton, MA: Smith College Libraries, 2005. The Cantina Press was the imprint of Professor Clarence Kennedy. This brief, 30-page publication has much useful information on Kennedy's academic career as an art historian, his interest and skill as a photographer, and his work as a typographer and printer.

Brasher, Milton E. *Rex Brasher, Painter of Birds.* New York: Rowman and Littlefield, 1961. This biography includes information (pp. 264–269) on Meriden's participation in the production of *Birds and Trees of North America.*

Bruhn, Thomas P. *Harold Hugo (1910–1985), Museum Patron.* Storrs, CT: The William Benton Museum of Art, 1989. Exhibition catalogue of drawings and graphic art given by Hugo to the Museum.

Chayt, Steven & Meryl. *Collotype: History, Practicum, Bibliography*. Winter Haven, FL: Anachronic Editions, 1983. A deluxe limited edition of 85 copies. It includes mention of the Meriden Gravure Co. but does not provide extensive information specific to the company.

Gardner, Warren F. (attributed to). "Harold Hugo Honored For Printing Craftsmanship." *Meriden Record*, July 6, 1954. An account of the party given for Harold at North Andover, MA, at which *Thomas Jefferson Among the Antiquities of Southern France in 1787* was issued.

Gardner, Warren F. "Harold Hugo Receives Yale Honorary Degree." *Meriden Record-Journal*, June 11, 1963.

Gardner, Warren F. "E. Harold Hugo, of Meriden Gravure." *Meriden Record-Journal*, September 11, 1985. Obituary article, unsigned but almost certainly written by Gardner.

Gardner, Warren F. "A Life Remembered at a Party He Asked his Friends to Enjoy." *Meriden Record-Journal*, December 10, 1985. An account of the memorial gathering held at Yale on December 6.

Goodspeed, George T. "E. Harold Hugo, 1910–1985." *Yearbook*, pp. 247–248. New York: The Century Association, 1986.

Hall, Elton W. *Harold Hugo as Sculptured by Leonard Baskin*. Boston: The Boston Athenaeum, 2008. Four-page illustrated brochure to mark the unveiling of Baskin's portrait sculpture of Hugo.

Hall, Elton W. "Hugo, E. Harold" entry in *American National Biography*, Volume 11, pp. 437–438. New York: Oxford University Press, 1999. A good, full-page summary of the work of Harold Hugo and The Meriden Gravure Co.

Hall, Elton W. *Printing as a Way of Life: Rocky Stinehour and the Stinehour Press*. Portland, Maine: The Baxter Society, 2014. Chapter 7 covers the entire history of the Meriden Gravure Co., in considerable detail.

Hidy, Lance. "The Mission and the Missionaries," pages 37–108 in *The SP Century*, edited by Scott-Martin Kosofsky. Boston: The Society of Printers & The Boston Public Library, 2006. A lengthy, informative and insightful account of the tradition of American fine printing as practiced in the northeast.

Hugo, Harold. *Gregg Anderson at the Meriden Gravure Company*. Meriden, CT: Privately Printed, 1946. This appeared in several versions: 12 pages and cover text only; 12 pages text plus 8 unnumbered illustration pages with a slightly different cover; 12 pages bound into a 1949 publication, printed for private circulation, *To Remember Gregg Anderson: Tributes by Members of The Columbiad Club, The Rounce and Coffin Club, The Roxburghe Club, The Zamorano Club.*

Hugo, Harold and Roderick D. Stinehour. *Vision & Revision, Introducing Meriden-Stinehour Inc.*, n.p., n.d. (1979). A photo essay to mark the merger of the Meriden Gravure Co and the Stinehour Press. A variant edition in paper covers was issued to honor the eighty-fourth birthday of Parker Allen.

Kelman, Eleanor. "Fine Press Printing: The Meriden Gravure Company." *Journal of the New Haven Colony Historical Society*, Vol. 31, No. 3, Summer 1985.

McLean, Ruari. "The Reproduction of Prints" in *Print Quarterly*, Vol. IV, No. 1. London, March 1987. Includes discussion of Meriden's quality control techniques and procedures.

McLean, Ruari. *True to Type: A Typographical Autobiography.* Oak Knoll Press, 2000. Includes information on Hugo and the Meriden Gravure Co.

McLean, Ruari. "Dr. Harold Hugo." *The Times* (London) September 12, 1985. An unsigned but lengthy and informative obituary.

Meriden City Directories, 1888-1978. New Haven CT: The Price & Lee Co. These directories, issued annually, are useful for determining the dates and place of residence and employment of people living or working in Meriden. They also contain advertisements and information on businesses in the city.

Peckham, John F. *Adventures in Printing: A talk on the Career of Harold Hugo Given at the Club of Odd Volumes.* Included in the publication is an Introductory Note by Rocky Stinehour, and a brief biographical summary by Peckham and the present author. Lunenburg Vermont, The Stinehour Press, 1995.

Reardon, Tom & Kent Kirby, "Collotype: Prince of the Printing Processes." *Printing History*, No. 25, Vol. XIV, No.1, 1991. Includes a brief survey of collotype at Meriden as well as other global practitioners of the craft.

Stulik, Dusan C. and Art Kaplan. *The Atlas of Analytical Signatures of Photographic Processes: Collotype.* Los Angeles: The Getty Conservation Institute, 2013. http://hdl.handle.net/10020/gci_pubs/atlas_analytical. A good, clear 32-page account of the technology and history of collotype printing in addition to technical information on identifying collotype and related processes.

Walker, Gay. *Harold Hugo & The Meriden Gravure Co.* 28 unnumbered pages. Privately printed by the author in an edition of fifty-five copies as

Columbiad Club Keepsake No. 111, April 1995. The text is a slight updating of a talk given at the memorial gathering for Harold held at Yale University, December 6, 1985. Much of this talk was taken from Ms. Walker's January 1985 William Addison Dwiggins Lecture, "Achievement in Excellence: The Meriden Gravure Co., & Harold Hugo."

Walker, Gay. "Printing for the United States: Meriden Gravure and the U.S. Government." *Printing History*, Vol. X, No. 2, 1988. A well-researched and detailed account of government contract printing done at Meriden before, during and after World War II.

Whitehill, Walter Muir, Julian P. Boyd and John F. Peckham. *A Portfolio Honoring Harold Hugo for His Contribution to Scholarly Printing.* n.p., 1978. Thirty-six fascicles, each with text and illustration contributed by scholarly institutions, plus front and back-matter fascicles. Whitehill wrote the Foreword, Boyd the Epilogue and Peckham the Memoir, "Harold at Meriden." Sinclair Hitchings edited for publication "Harold Hugo on the Meriden Gravure Company," a series of tape-recorded interviews done in 1964.

Whitehill, Walter Muir, *Fred Anthoensen, A Lecture Given at The Composing Room, New York City, 23 February 1966*, Portland, ME: The Anthoensen Press, n.d. (1966). Includes as an epilogue a two-page letter by Paul Standard to Anthoensen, reproduced in Standard's fine Italic hand, in praise of Anthoensen. It was reprinted yet again (without the Paul Standard epilogue) as Chapter XIV of *Analecta Biographica: A Handful of New England Portraits*, Brattleboro, Vermont: The Stephen Greene Press, 1969.

Wright, Helena E. *Imperishable Beauty: Pictures Printed in Collotype.* Washington: National Museum of American History, 1988. Catalogue of an exhibition held in the Hall of Printing and Graphic Arts at the Smithsonian Institution. An illustrated history of collotype as practiced in Europe and America, with mention of Meriden Gravure.

Wright, Helena E. "Bierstadt and the Business of Printmaking." See Anderson and Ferber, *Albert Bierstadt: Art & Enterprise.* pp. 267–294.

Index

This book has been printed in an edition of 500 copies.
The type is Minion, designed by Robert Slimbach in 1990.